Aircraft Engine Test Beds

British Jet Fighters and Bombers

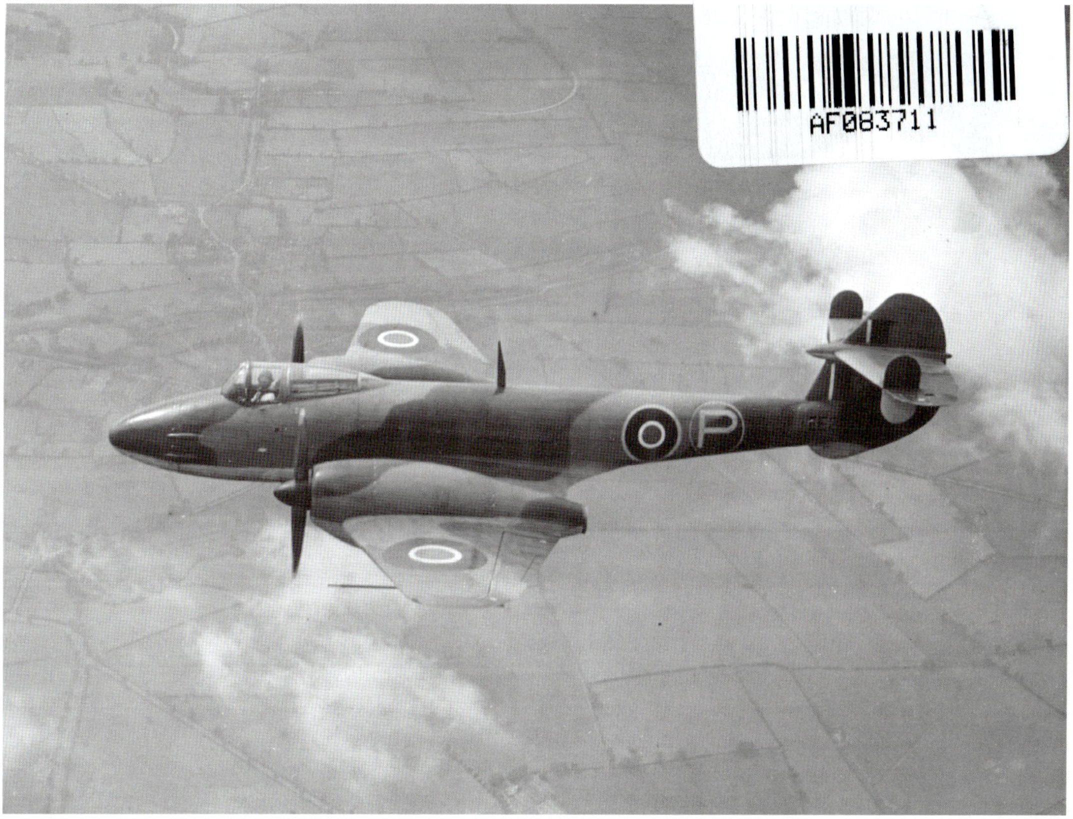

TONY BUTTLER, AMRAeS

KEY Books

Front cover image: RA490 pictured with its Nene engines newly fitted. (National Aerospace Library)

Contents page image: When this photo was made XA902 had received Spey engines in its inner engine nacelles. (Rolls-Royce)

Published by Key Books
An imprint of Key Publishing Ltd
PO Box 100
Stamford
Lincs PE19 1XQ

www.keypublishing.com

The rights of Tony Buttler, AMRAeS to be identified as the author of this book has been asserted in accordance with the Copyright, Designs and Patents Act 1988 Sections 77 and 78.

Copyright © Tony Buttler, AMRAeS, 2022

ISBN 978 1 80282 248 9

All rights reserved. Reproduction in whole or in part in any form whatsoever or by any means is strictly prohibited without the prior permission of the Publisher.

Typeset by SJmagic DESIGN SERVICES, India.

Contents

Introduction and Acknowledgements 4
Chapter 1   Gloster Meteor 6
Chapter 2   De Havilland Vampire 34
Chapter 3   Hawker Aircraft 45
Chapter 4   Gloster Javelin 55
Chapter 5   English Electric Canberra 66
Chapter 6   Short S.A.4 Sperrin and Vickers Valiant 80
Chapter 7   Avro Vulcan 87
Glossary and Bibliography 96

Introduction and Acknowledgements

Over the last few years, I have had the good fortune to write several articles for *Aeroplane* and *Aeromilitaria* magazines which covered those examples of British jet fighters and bombers that were fitted with new, different or additional engines, both jet and rocket, to enable them to serve as test beds for these powerplants. In addition, a book I had published with The History Press on the Hawker Hunter included a brief section on the examples of this fighter type that were used for the same purpose. It became clear that a modest revising, expanding and bringing together of these items, to show just how much gas turbine trials work was undertaken in the UK by 'jet fighter and bomber airframes', might be a worthwhile exercise. You have the result.

Plenty has been written about the numerous British two- and four-engine piston bombers that were used as test beds for the first generations of British jet engines (the Vickers Wellington, Avro Lancaster and Lincoln and the civil variant Lancastrian and Lincolnian). The same can be said for the Meteor jet fighter and for the Vulcan and Canberra jet bomber test beds. However, in my opinion, the Vampire, Javelin and Hunter have perhaps gone a little under the radar. The prototype Sperrin's specific role as a test bed and the relatively small contribution by the Valiant have, I think, also tended to be overlooked. Hopefully this volume will bring a little more balance to proceedings.

Not all home-grown British jet fighters and bombers had examples used as engine test beds – as far as I know, no production examples of the de Havilland Venom and Sea Vixen, English Electric Lightning, Hawker Seahawk and Supermarine Attacker, Swift and Scimitar, or the Handley Page Victor V-Bomber, were ever used to test new engines, that is beyond new versions of the powerplants already used by these types. Some of course did find employment in other research roles such as new weapons, and the English Electric P.1A did fly with Armstrong Siddeley Sapphires when production P.1B Lightnings had Rolls-Royce Avons. So, we are left with the de Havilland Vampire, Gloster Meteor and Javelin, Hawker Hunter and P.1072 (the latter of which was an adaptation of the P.1040 prototype that led to the Seahawk), the English Electric Canberra, Short Sperrin and Vickers Valiant, and bringing up the rear the Avro Vulcan.

All of the main jet fighter and bomber airframes known to have participated in this field of work have been included. The book concentrates primarily (but not exclusively) on airframes that received engines for assessment for potential use in other types, or which were used to test new features like reheat or thrust reversal. At times this produced odd and quite remarkable-looking airframes. However, in some cases engines were fitted in an effort to improve an aircraft type's performance – for example the Rolls-Royce Nene used in the Vampire. In general, the photographic coverage (with a few exceptions) is excellent, though in many cases the only public performances for the majority of these aircraft came at the annual Society of British Aircraft Constructors (SBAC) Show at Farnborough. There is undoubtedly great interest within the historical aviation community for material on prototypes and experimental aeroplanes and I hope that this volume will provide a useful reference for this particular category of research type, and also fill a small gap in the documented coverage. It has been a fascinating and hugely enjoyable task putting this material into a book, as was the case when my research for the original articles began well over a decade ago.

I am indebted in particular to the late David Birch at Rolls-Royce Heritage Trust at Derby, to *Aeroplane* editor Ben Dunnell for first commissioning and accepting some of these pieces in their original form, the Farnborough Air Sciences Trust (FAST), Chris Farara of the Brooklands Museum's Hawker archive, Barry Guess of BAE Systems Archive, Farnborough, Bob Hercock of Rolls-Royce Heritage at Filton (in particular for de Havilland Engines records and photos), Tim Kershaw of the Jet Age Museum, Victoria Northridge at Coventry Archives, Terry Panopalis for allowing me to use some tremendous photos, Amy Rigg at The History Press, and the National Archives at Kew.

Special thanks, as ever, go to my great friends Phil Butler, for record card information for individual airframe histories and for photos, and to Chris Gibson, who first commissioned several pieces when he was editor of Air-Britain's *Aeromilitaria* magazine. I am most grateful to you all!

Tony Buttler MA, AMRAeS
Bretforton, November 2021

Jet Deflection Meteor RA490 takes off during the Farnborough Jubilee event on 9 July 1965.

Chapter 1
Gloster Meteor

Such was the nature of the new gas turbine (jet engine) that, apart from the pure research aircraft built at the start to test it (the famous Gloster E.28/39 of 1941), there was a need to evaluate the new power units in the air in airframes that were normally driven by safe and reliable piston engines. Initially, a Vickers Wellington bomber was used but, as more engine types were developed during the years following the end of World War Two, examples of the larger Avro Lancaster and Lincoln also found employment in this role. However, the performance of these heavy early 1940s technology airframes was limited; something faster was soon required and in due course established jet fighters and bombers were absorbed into the various programmes.

The Gloster Meteor, Britain's first jet fighter, was powered by two jet engines mounted in wing nacelles and this configuration made the airframe convenient, indeed ideal, to test alternative units. The first of eight prototypes (serial DG206) made the type's maiden flight on 5 March 1943 but this aircraft, and also DG207, had de Havilland engines installed because the prototypes fitted with Frank Whittle/Power Jets-designed engines were not yet flight ready. The breakdown of Meteor prototypes is as follows, and all of them can perhaps be considered as engine test beds since the concept of jet-powered flight was so new and advanced.

DG202 – Rover-built W.2B/23;
DG203 – Power Jets W.2/500 and later W.2/700;
DG204 – Metropolitan-Vickers F.2;
DG205 – Rover W.2B;
DG206 – de Havilland H.1 (which became the Goblin);
DG207 – de Havilland H.1b;
DG208 – Rolls-Royce W.2B/23 (became the RB.23 Welland);
DG209 – Rolls-Royce W.2B/37 (became the RB.37 Derwent).

The Sapphire Meteor WA820 in a picture taken by Gloster photographer Russell Adams, most likely on 17 August 1951. The photographer was in a two-seat Meteor T.7 flown by company test pilot Brian Smith. (Jet Age Museum)

DG204

This prototype was fitted with Metro-Vick F.2 axial jets and, although higher power ratings were expected, at this stage the take-off thrust of each engine was nominally 1,800lb with (excluding the radio and its mounting) the prototype's all-up-weight with 300gal of fuel totalling 12,190lb. DG204 was moved by road to the Royal Aircraft Establishment (RAE) Farnborough and Wg Cdr H. J. Wilson took the aircraft on an 11-minute first flight on 13 November 1943. A second flight (with replacement engines) followed on 19 December, another on the 22nd, four on the 25th (Christmas Day!) and more were planned for 4 January 1944. These sorties were shared by Wilson and by Sqn Ldr W. D. B. S. Davie and during the week before Christmas DG204 achieved a speed of approximately 373mph at 15,000ft.

Tragically, during the second sortie on 4 January the compressor in DG204's port engine burst (through over-speeding), the end of the tail structure failed and broke off in the air and the aircraft crashed near Farnborough. Ejection seats were not yet available to the pilots of high-performance aircraft and Davie died when he was hit by the aircraft after having baled out. There was a clear blue sky that day and many Farnborough staff, enjoying a lunch break, could see DG204's exhaust as it flew overhead. Suddenly, they saw the Meteor perform large gyrations in its flight path and then break apart, and the detached tail actually landed on the roof of the RAE Foundry, close to the engine test beds. When it disintegrated the compressor drum sliced through the airframe, although the fuselage had itself failed in an unusual way that necessitated special strength tests on the fuselage and tail of another Meteor.

Photos of the Metropolitan-Vickers F.2-powered Meteor prototype DG204/G taken in July 1943. 'G' stood for 'Guard' while on the ground during the war, such was the secrecy surrounding the jet at this time. These were taken at the Gloster experimental department at Bentham prior to despatch for Farnborough for the first flight and show the underslung nacelles housing the first British axial-flow jet engines to fly. (Crown Copyright)

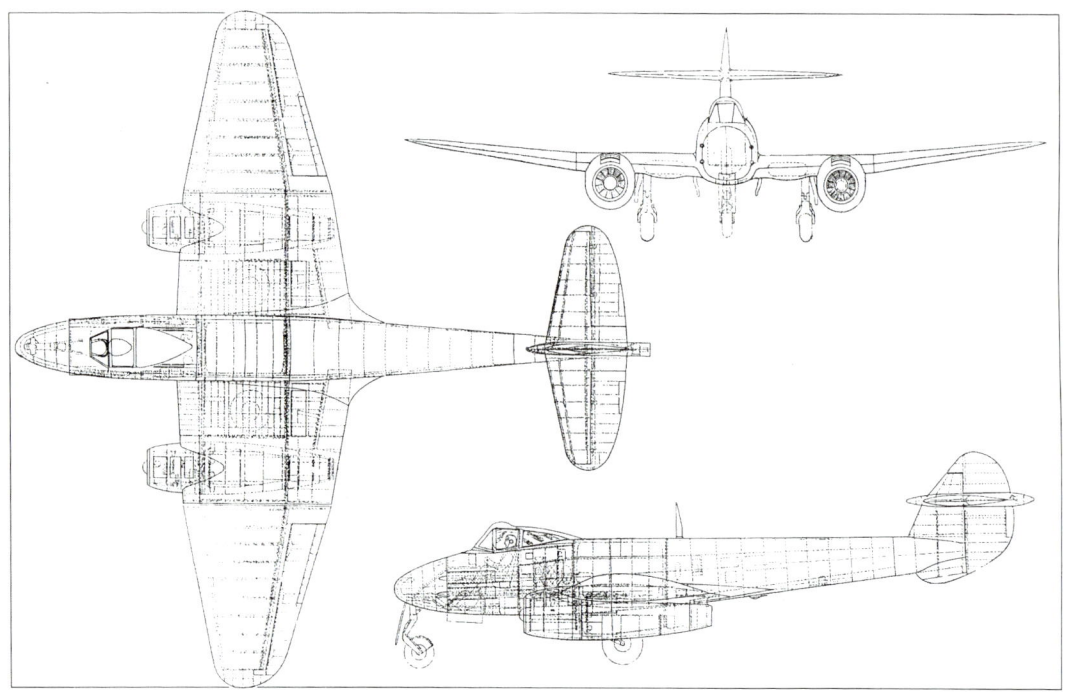

Above: Three-view drawing of DG204.

Left: This picture of DG204/G was taken at Farnborough on 20 November 1943, just one week after the maiden flight. (Farnborough Air Sciences Trust)

The objective of DG204's trials had been to obtain engine performance data, but the crash brought these tests to an early conclusion and it appears only nine flights were completed within the programme. These had included level speeds at 3,000ft, 15,000ft and 20,000ft and covering a range of engine speeds, plus just one reliable climb to 15,000ft. The time taken to reach 15,085ft was 8 minutes and 14 seconds, the mean rate of climb being 1,830ft/min. DG204 had been considered rather 'rough' to fly because its controls were heavy, and at 20,000ft the pilot reported that the Meteor 'wallowed' during level flight (a term describing a combination of rolling and low frequency 'snaking'). RAE also noted that the noise generated by DG204 in the air was less than for other jet propulsion aeroplanes. Criticism was made of the difficulty of getting out of a Meteor in an emergency, but an ejection seat was not introduced to production machines until the arrival of the Meteor F.Mk.8 in 1948. A Ministry of Supply document of 28 October 1947 indicated that a follow-up aircraft might be transferred to test F2/4 units (see RA490 below).

Splendid photo of the F.2 installation with the covers removed for DG204/G's port side. (National Aerospace Library)

DG206

The first flight-ready Meteor prototype had been DG206, the de Havilland H.1 test bed, and Gloster test pilot Michael Daunt performed the first flight, from RAF Cranwell, on 5 March 1943. Further flights were made from Newmarket and from Barford St. John and by 24 September the Meteor had accumulated a total flight time of 10.5 hours. On 19 December DG206 arrived at RAE Farnborough by road and, with new engines installed, Wg Cdr Wilson flew it on 2 March 1944. He completed five more sorties during March and May, but after that DG206 appears to have been little used.

The Halford H.1-powered DG206/G, the first Meteor to fly, is seen at Farnborough with a test bed Lancaster to the left. The H.1 (Goblin) engine had a larger diameter than the Whittle W.2B units in other Meteors and so required a larger 'banjo' to carry the wing spar loads, which increased the span slightly and made the nacelle fatter. (Crown Copyright)

DG207

This airframe was earmarked as the Meteor Mk.II prototype, to be fitted with H.1b (Goblin) engines, but in August 1944 the Mk.II production run was postponed and eventually cancelled because higher priority went to Goblin production for the de Havilland Vampire. Consequently, this airframe did not fly (from Moreton Valence) until 24 July 1945 with Gloster test pilot John Grierson in the cockpit. At one stage there were plans to fit two 5,000lb thrust Nene units in DG207, but instead it was earmarked for further Goblin development flying. Based at Hatfield from 6 September it was, however, little used, because DG207's endurance was insufficient to permit enough flight time at high altitude. On 5 November 1948 DG207 was taken by road to RAF Locking to serve as ground instructional airframe 6591M with No 5 School of Technical Training, and in April 1955 it was transferred to No 1 School of Technical Training at RAF Halton.

Power Jets W.2/700

Prototype DG203, alongside several early production Meteors, were all used to test the W.2/700 engine earmarked for the Miles M.52 supersonic research aircraft (which was later cancelled). DG203 first flew with W.2/700s in October 1944, and F.Mk.I EE211 began flying from RAE Farnborough with 2,000lb thrust W.2/700s in March 1945. Next, EE215 first flew with this engine on 18 October 1945, and from May 1946 it was used for reheat and speed experiments flying out of Armstrong Whitworth's Bitteswell airfield. From February 1945 EE221 flew reheat and fuel consumption flight trials with the W2/700 before joining the National Gas Turbine Establishment (NGTE) at Pystock. A Ministry report noted that tests with EE221, a standard airframe with four cannon, gave a top speed of 443mph at ground level and 465mph at 25,000ft, and a maximum rate of climb of 4,200ft/min at ground level.

Meteor Mk I EE211 pictured with W.2/700s installed in long nacelles.

Trent Meteor

As the jet matured, further lines of development looked at using the new power source in other ways. One was the turboprop, where a propeller was fitted in front of the jet, and Meteor Mk.I EE227 became the first 'prop-jet' aircraft in the world to fly – specifically the first solely by aero turbines driving propellers. The engine in question was the Rolls-Royce RB.50, an adaptation of the firm's Derwent turbojet fitted with a gearbox to drive a five-blade airscrew (propeller) and which in 1945 was named Trent. This gave 1,000lb of thrust and 800hp, the total power coming from a combination of jet pipe thrust and propeller forces. Ballast was placed inside EE227's nose to replace the guns and ammunition, but it proved impossible to get the weight under 13,865lb. In addition, the undercarriage was extended by 6in to provide propeller clearance and owing to the different thrust line and the effect of the slipstream from the propellers, the tailplane incidence was increased by one degree. There was no intention of putting this type into production, it was purely experimental.

Consideration was given to converting two airframes, but in the end only EE227 was so treated. This aircraft had served with 616 Squadron before going to RAE Farnborough for flight stability tests with the upper part of its fin removed. In March 1945 it went to Hucknall for the Trent conversion, and on completion was then taken to RAF Church Broughton which, with hard runways, was the base for Rolls-Royce's experimental jet flying (the Jet Propulsion Flight Development Section). It was from here on 20 September 1945 that EE227 made a first flight with its turboprop powerplant, crewed by Gloster test pilot Eric Greenwood. The aircraft displayed the shortest take-off yet seen by a Meteor, but serious

Above left and above right: Three-quarter angle images of the Trent Meteor EE227/G showing the original five-blade propellers (or using the terminology of the time, airscrews) and then the set with cropped blades.

Below: Rolls-Royce photo of the attractive-looking Trent Meteor during a test flight. (Rolls-Royce)

This set of views of EE227 were all apparently taken on 9 July 1947. Two particular features of the turboprop Meteor, the extremely neat nacelles and the small finlets, are well shown. (Rolls-Royce)

handling problems were apparent immediately. Consequently, EE227 went back to Gloster's Moreton Valence facility for seven months for these problems to be investigated, and during this period auxiliary fins were fitted to the horizontal tailplane.

Shortly after returning to Church Broughton, in May 1946 this field was closed to flying and all of Rolls-Royce's jet work was moved to Hucknall's grass field. However, the Trent Meteor was not permitted to operate from grass, so in October it was transferred to Bitteswell which had a hard runway. The handling problems were still prevalent – in particular, severe yawing when the throttle was closed and made worse by a severe nose-down pitch. It was considered that the propellers might be the cause of these poor characteristics and so the original 7ft 7in diameter set was replaced by a new blade cropped to just 4ft 10in diameter. As such, the propellers absorbed less power and in consequence the thrust output was increased to 1,400lb and 350hp. However, 25 hours of flying in this configuration gave no improvement, so the original propeller was restored for the rest of EE227's turboprop career. It was finally established that the handling problems were caused by the interaction between throttle and propeller, and in 1947 the Trent Meteor's engine controls were revised so that just one lever could operate both the throttle and the propeller speed. As a result, the handling was greatly improved, to a level on a par with a jet-powered Meteor Mk.III. The violent and dangerous changes of trim previously experienced had gone completely, as were symptoms of a fake stall at certain speeds.

On completion of the Trent programme, totalling 147 hours flight time and mostly conducted by Rolls-Royce test pilot Wg Cdr Andrew McDowall, in February and March 1948 several other test pilots, from the Aeroplane and Armament Experimental Establishment (A&AEE) Boscombe Down, RAE Farnborough, NGTE Pystock and six airframe and engine manufacturers, flew EE227 to gain turboprop experience. The aircraft was then converted back to standard at Hucknall with Derwent engines for ground-running tests, prior to being dismantled and despatched to Farnborough. The Trent units themselves had experienced few mechanical problems and the maximum recorded speed had been 440mph at 10,000ft.

One pilot given the opportunity to experience turboprop flying in EE227 was *Aeroplane* magazine's Richard Worcester and his report appeared in the 26 March 1948 issue. He noted that:

On opening the throttles fully, the initial acceleration is tremendous and far greater than [for] pure jets; it is indeed comparable with piston engines in the largest sizes. The elevator control seems very heavy and a strong pull-force is needed to start the climb. In the air the aileron control is pleasantly easy. The rudder is reasonably sensitive, and the elevator is the hardest of the three. Under asymmetric power the aircraft will fly with negligible opposite rudder and no off-setting rudder trim appears to be necessary. The engines can be suddenly closed to the idling position with no ill-effects whatever as the blades at once coarsen pitch to reduce airscrew-disc drag. When fully opened, the response is instantaneous and precisely like that of a piston engine; the acceleration from say 150mph to 250mph is most marked. The safety speed is 145mph, and if an engine cuts on take-off it is necessary to close the throttle of the dead engine to reduce blade drag by getting into coarse pitch at once.

Worcester added:

In the air the noise is not appreciably more than on some pure jet aircraft and is considerably less than a piston-engined aircraft of comparable performance. The actual performance of the Trent Meteor is not of great significance since the object has been to develop the prop-jet engine. Nevertheless, the Trent Meteor is as fast as the most highly developed piston-engined interceptor fighters [and] it is faster than the earlier Meteors with the original Derwent engines. The aircraft lands naturally in a tail-down position with the nosewheel well off the ground. With the high initial speed, the braking effect of the airscrews is noticeable.

Reheat

Three Meteor Mk.IV airframes were used for afterburning (reheat) trials, although reheat was never used on any Service aeroplanes. The first was serial RA435, which was delivered to Rolls-Royce for Derwent V development work on 28 August 1947. For its reheat trials the Meteor was fitted with simple extended rear longitudinal nacelles, inside which were 18.9in diameter jetpipes instead of the original 16in format. The reheat system increased RA435's weight by 980lb up to 15,630lb, though 400lb of this was ballast used to control the centre of gravity (CofG). With reheat unlit the original installation had a lower basic thrust, but this was dealt with by fitting variable nozzles.

RA435 first flew after modification on 10 June 1949, although the reheat was not used in flight until later that month. The programme looked at engine handling, engine and associated structure temperatures, and relighting. A public performance followed at the SBAC Show at Farnborough in

Above: The Derwent reheat installation and extended nacelles on RA435.

Below: Apart from the reheat fitting there were relatively few changes to RA435's airframe.

Below: RA435 on the runway at Farnborough in September 1949. The noise created by afterburning, well known to airshow spectators today of course, must have been a big surprise to those new to it in the late 1940s.

Reheat trials Meteor VT196 pictured at Hucknall. (Roll-Royce Heritage Trust, Derby)

Rare air-to-air photo of VT196, which has enough detail to show how the nacelle and jetpipe were of a more streamlined form than those previously used on RA435. (Rolls-Royce)

September, the first time that reheat had been demonstrated to the public. *Flight* magazine reported on the display performed by Capt Shepherd. Having made flypasts at 'moderate' speeds, Shepherd lit the 'reheaters' and the 'Meteor thunderously ascended to high heaven, boring a clean hole through a cloud'. The aircraft was described as 'painfully noisy' and smelt 'offensively', but the sight of it 'receding sunwards until it becomes a mere pin-point of light' was something to remember.

In its early form RA435's reheat could stay lit to a height of only 22,000ft, but improvements to the fuel system increased this to 41,000ft. In all, 28.5 hours of airtime were recorded with reheat lit for 3.5 hours, but the system was temperamental and on 23 occasions the reheat failed to light

(119 successful relights were recorded). If an engine failed in the air there was a pronounced swing, and some take-offs were also made with reheat operating in the port engine only. Here, the take-off started with the aircraft aiming at the hangers before the curved run (from the uneven thrust) brought it to the correct direction for lift-off. After a year-long flying programme RA435 was replaced in 1950 by VT196 and VZ608 with more streamlined nozzles, which could be both opened and closed during flight. In December 1951 RA435 became instructional airframe 7131M at RAF St Athan.

VT196 was used by Rolls-Royce Hucknall for Derwent V development from July 1948, before being fitted with afterburners for basic research into reheat technology (this modification increased the engine's thrust to 4,400lb). In July 1953 it was sent to Canada for trials with the Schmidt afterburner, as such on loan to the Canadian National Research Council (CNRC), though officially on charge with the Royal Canadian Air Force. It is understood that the CNRC trials formed part of the development of afterburners for the Avro Orenda axial-flow turbojet. After these trials were completed in June 1955, in mid-1956 VT196 was returned to England and converted by Flight Refuelling Ltd into a U.Mk.15 drone. VZ608's career is discussed later.

Beryl and Jet Deflection

Originally an F.Mk.IV/4, RA490 was refitted with two 3,950lb thrust Metropolitan-Vickers F.2 Beryl axial-flow turbojets. The engines were in fact the F.2 Series 4 developed from the version flown in DG204. Underslung nacelles had been used to accommodate the power units on that aircraft because there was insufficient room to fit them between the Meteor's characteristic 'banjo'-type wing spars.

Pictures of RA490 which show the larger nacelles required to accommodate the Beryl engines. (Jet Age Museum)

Beryl Meteor RA490 attended the 1948 Farnborough Show. (Phil Butler)

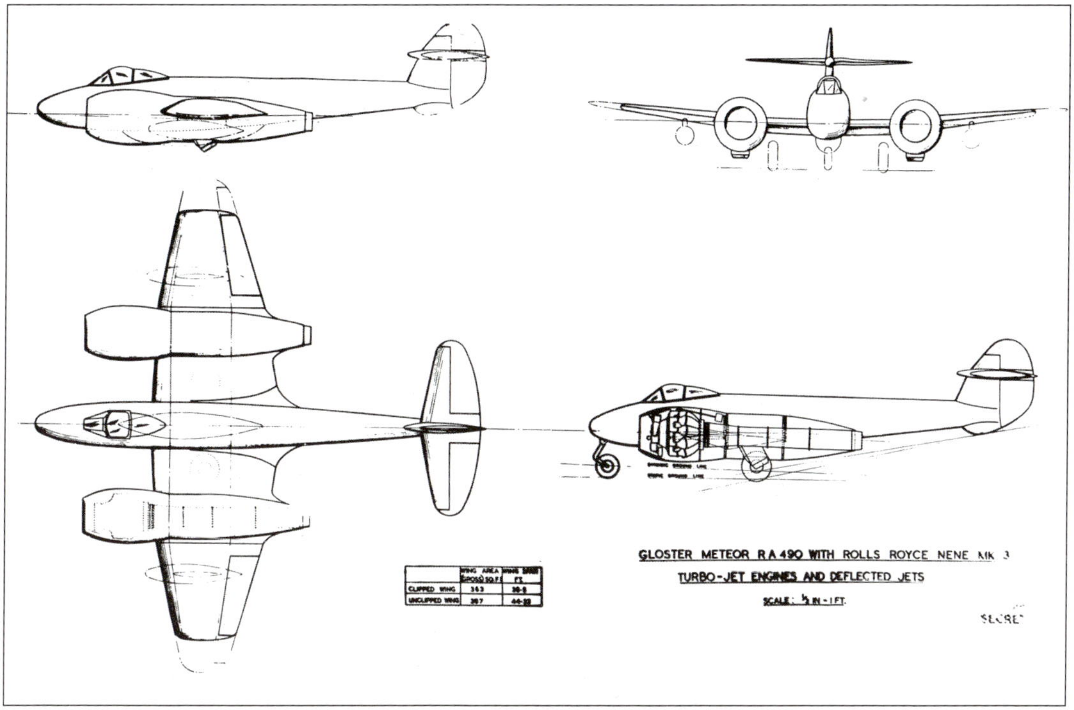

This original Westland drawing shows the first proposal for converting RA490 for jet deflection experiments. The aircraft still has its Mk 4 rear fuselage and tail, but as built the nacelles were of a different shape. (Westland Archive)

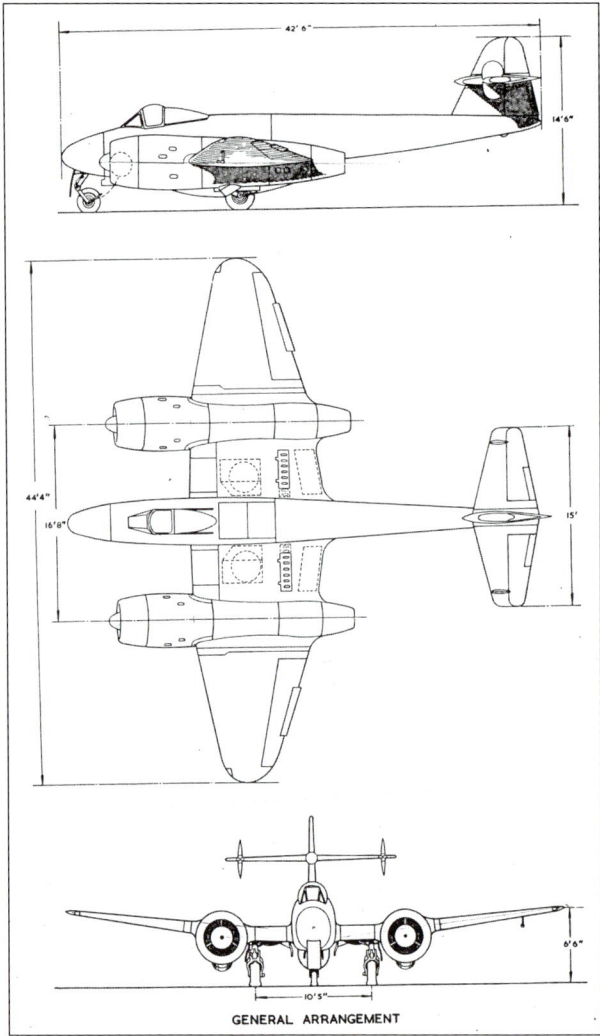

Left: Three-view drawing of the Jet Deflection Meteor as completed, with a Mk 8 rear fuselage and fin. (Westland Archive)

Below: Airframe and engine detail for the jet deflection test bed. (Westland Archive)

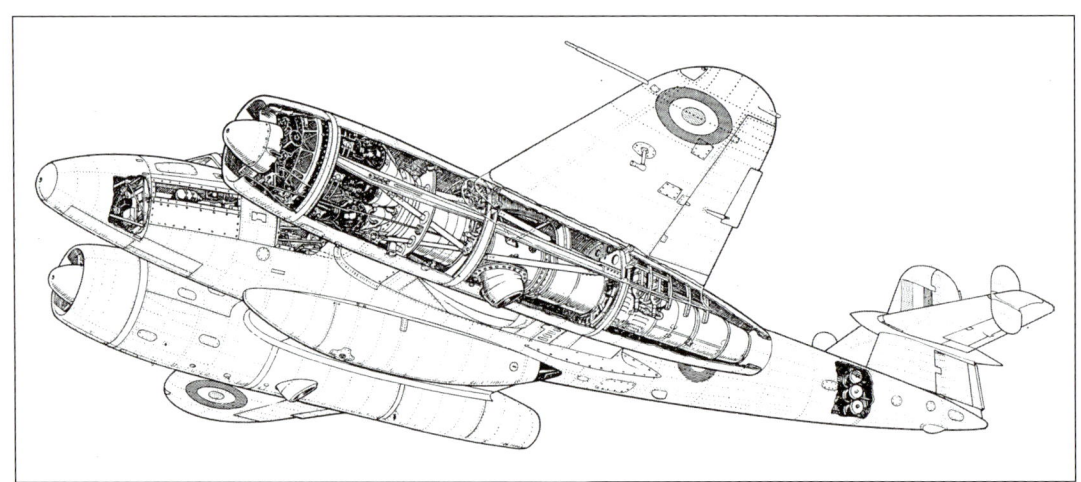

However, RA490 had a modified wing centre with inverted reinforced U-sections in both front and rear spars, which permitted different engines to be accommodated with sufficient ground clearance. RA490 was handed over to Gloster on 2 December 1947 and the Beryls were installed in 1948. The new nacelles were longer than for a standard Meteor and of basically cylindrical form, and with the gun ports faired over this very clean airframe was very fast.

It could also climb quickly and just before Christmas 1948 the press reported that RA490 had, in October at Moreton Valence, recorded a time to 10,000ft of 55 seconds, to 15,000ft of 1 minute and 35 seconds, 20,000ft in 2 minutes and 5 seconds, 25,000ft in 2 minutes 50 seconds, 30,000ft in 3 minutes 43 seconds, 35,000ft in 5 minutes 55 seconds, and 40,000ft in 7 minutes 31 seconds. The altitude sorties were flown by Gloster's famous test pilot Sqn Ldr Jan Żurakowski. The Beryl Meteor had an all-up weight of 15,455lb, some 255lb more than a standard Derwent-powered Mk.4, but the new engines had taken RA490 up to 30,000ft and 40,000ft in less than half the time needed by a Mk.4 (the latter's figures were 8 minutes and 17 minutes, respectively). RA490 attended the SBAC Show held at Farnborough

RA490 pictured with its Nene engines newly fitted. (National Aerospace Library)

Above left: Close-up of the downward nozzle from one of RA490's deflection boxes. The deflector jetpipe protruded underneath the engine nacelle at approximately the mid-way point. (Westland)

Above right: Finlets were added to provide RA490 with adequate control at the lower flying speeds obtained with jet deflection. (Westland)

in September 1948. Next, on 11 January 1949 it went from Gloster to the NGTE at Bitteswell for Beryl engine development work, the loan for these trials being extended to mid-December 1951. Then in July 1952 RA490 was allotted to Westland Aircraft for experiments in jet deflection.

The introduction of high-performance jet aircraft to the world's air arms raised several issues, one of which was to keep their landing and stalling speeds sufficiently low to ensure adequate safety. As the maximum speed went up it became more difficult to match that with the ability to fly slowly enough to ensure that a safe landing could be affected. One potential solution was to create 'artificial' lift to reduce the stalling speed, to produce some upwards thrust by deflecting an aeroplane's jet efflux downwards at an angle to the line of flight. In fact, jet deflection was considered especially promising, particularly for applications where orthodox high-lift devices would prove inadequate. To begin, the additional lift provided by deflected jets would be substantially independent of the aircraft's forward speed.

For the RAE Golden Jubilee celebrations in July 1955 RA490 was displayed outside Farnborough's 'A' Shed. The size of the guards covering the Nene nacelle intakes was rather larger than those needed by a normal Meteor.

This photo shows two heavily modified Meteors during their display at the Farnborough Jubilee; the jet deflection RA490 and then serial WK935 which had been fitted with an extended nose and second prone-pilot cockpit. The date is 9 July 1955. (Martin Richmond)

The deflection of some or all of the jet thrust downwards would also make it possible to reduce the effective weight on the wing by a corresponding figure, which as stated would reduce the stalling speed. For the pilot it would be easier to make an approach without having to make changes in airspeed or attitude. In addition, the aircraft's rate of descent could be checked quickly (or indeed converted into a climb) without having to increase the forward speed by very much. Finally, vertical acceleration could be applied immediately simply through moving the engine power lever forward and the installation was so arranged that no change of trim would result.

This programme was not pure vertical take-off research using direct-lift, but rather it was intended to help in the design and operation of very fast aircraft of conventional configuration. The requirements were established by a joint NGTE/RAE group which stipulated in particular that the machine had to be capable of climbing at an angle of 5 degrees with the jet deflection in operation. The deflection angle eventually chosen was 63 degrees – at no point was a full 90 degrees of deflection considered because it was desired to retain a component of forward thrust throughout.

RA490's jet deflection installation was two 5,000lb Nenes. The deflectors themselves were designed by NGTE and the nacelles were massive, because it was essential to keep the downward-directing jet orifices (the deflected-thrust line) close to the CofG. To achieve this the engines had to be moved forward by some distance, in fact ahead of the wing spar entirely. In the end, the nacelles stretched more than 8ft beyond the wing and the deflection nozzles protruded out of the nacelle underside at a position just to the rear of the wing leading edge. The nozzles themselves were very short but, even with both tyres and oleos fully compressed, the forward edges of the jetpipes were just an inch or so clear of the runway. To provide deflection a vane placed inside one of the internal sections could be redirected to divert the exhaust gases downwards through a deflector box mounted between the main wing spars to the jetpipes. Each Nene also had a normal conventional straight-through jetpipe that stretched rearwards from the deflector to extend through the spar 'banjo'.

The jet deflector added 410lb in weight. To counteract the considerable nose heaviness (despite guns, ammunition and armour having been removed) ballast had to go inside the aircraft's rear fuselage. In addition, to counteract any lateral instability brought about by the increased side area of the nacelles, a Mk.8-type rear fuselage and tail were fitted instead of the older Mk.4 form. Additional Trent Meteor-style outer finlets were also introduced to increase the directional stability at low speeds but RA490's original Mk.4 nose and centre section were retained, though a strengthened Meteor NF.11 nosewheel was added. Finally, outer wings normally used by the PR.Mk.10 were fitted to retain lateral control if an engine failed. This gave a span of 44ft 4in, the biggest flown by any Meteor. The outcome was a quite complex hybrid aircraft. The conversion was undertaken at Westland, apparently from July 1953 onwards. Pre-flight ground running at the firm's Merryfield airfield uncovered few problems and RA490 made its first flight in this new form on 15 May 1954, from Boscombe Down with Westland chief test pilot Sqn Ldr Leo de Vigne in the cockpit. It was the first British aeroplane to employ jet deflection.

Flying out of Boscombe and Merryfield, de Vigne accumulated over 15 flight hours in RA490 and at this stage all take-offs and landings were made conventionally. Deflection was not employed at heights lower than 3,000ft and at that altitude the Nenes produced 88 per cent of their sea level thrust at full power. In flight the jets themselves were set at either 'all horizontal' or 'all deflected'. De Vigne was able to report that the deflection of the jet stream reduced the stalling speed considerably, and that Westland's engineers had been so successful in eliminating trim changes that he had found it impossible in flight to tell whether the jets were deflected or not without having to check the cockpit instruments.

On 16 August 1954 RA490 was transferred to Farnborough. RAE's pilots found that when it was flown as a conventional aeroplane, its low speed and stalling behaviour were pretty similar to a normal Meteor. When the deflected jets were operating, the lower speeds that could be achieved were dependent of course on the volume of thrust deflected while, at a given engine speed, any increase in the ambient temperature and/or the flying altitude produced a noticeable loss of performance. Using jet deflection the lowest indicated airspeed ever recorded was 65 knots using full power (only ever achieved by one pilot), but a more consistently recorded figure (in normal winter conditions and at 2,000ft altitude) was 75 knots.

A normal approach using the deflected jets would be made at 120 knots forward speed. When the height came down to 200ft the defection of the jet stream would be activated with the forward power reduced, but as the forward speed fell away power would be applied once more. The stalling speed with a clean aircraft and no deflection was 110 knots, and then with flaps up, undercarriage down, jets deflected and 60 per cent engine power, it was 95 knots. At these low speeds the control about all three axes was satisfactory with adequate aileron and rudder response (though the aileron control tended to become sluggish), but the elevator response was less good and was aggravated by the pilot feeling buffet on his control stick.

In March 1956 RA490 was allotted to RAE Bedford (Thurleigh) under a year-long loan and delivered on 27 April. Here it was used for numerous trials and several pilots took the opportunity of sampling flight using deflected jet power. RA490 also returned to Farnborough to take part in an event to mark the RAE's Golden Jubilee. This was held from 7 to 9 July 1955 and was open to press and industry, but not to the public. During the flying display, RA490 flew in a unique formation of three unique aeroplanes – the jet-deflection Meteor, the Boulton Paul P.111 delta wing research aircraft and another Meteor (WK935) fitted with a nose extension and second cockpit for prone-pilot research. Contemporary press accounts state how the delta 'touched down decidedly "hot" and dropped its tail parachute; whereupon the jet-deflection Meteor, sitting down amazingly slowly, blasted the 'chute neatly off the runway'.

After completing its trials RA490 was Struck off Charge on 12 April 1957 and used for firefighting training at Thurleigh. Overall, the Nene jet deflection installation had proved very successful and it was calculated that, at the 'normal' minimum speed of 75 knots, the deflected thrust was supporting over 40 per cent of the aircraft's weight. The experiment had shown that the Meteor's stability and control were (for test flying purposes) adequate at speeds down to 70 per cent of the normal (power-off) stalling speed, and when flown conventionally RA490's low-speed performance still came near normal Meteor figures. The test bed had proved the complete practicability of the jet deflection system but the concept was never generally adopted worldwide, primarily because of the penalty it brought in weight. Other methods were developed to provide high-speed aircraft with the ability to fly slowly enough to be able to land safely. For example, internal blown flaps were used by some carrier- and land-based fast jets to improve their low-speed lift for take-off and landing.

Avon and Atar

On 18 November 1948 Mk.4 RA491 arrived at Rolls-Royce to have two 6,000lb Avon RA.2s installed in a modification that required the centre wing to be strengthened to take nacelles of considerable size, 25 per cent larger than usual (the aircraft had been allotted for conversion in November 1947). Gloster's Sqn Ldr Bill Waterton took RA491 on its maiden flight on 29 April 1949 from Moreton Valance as the first aircraft to be powered entirely by Avon engines, and it was also the most powerful Meteor to fly so far. On 9 June it went to Rolls-Royce and improved 6,500lb RA3s were installed in April 1950. With these RA491 reached 40,000ft in just 2.7 minutes and 50,000ft in 3.65 minutes, and much of the test flying was undertaken at over 45,000ft to assess the engines and their handling at high altitudes.

The aircraft was shown at the 1949 and 1950 SBAC Displays where it demonstrated this extraordinary rate of climb. At the 1949 Farnborough Airshow, *Flight* noted that this was the 'world's most powerful and fastest-climbing single-seater, adapted for two of the still-secret axial Avons'. Flown at the event by Rolls test pilot Sqn Ldr Jim Heyworth, RA491 reportedly 'rocketed up into the blue and plummeted down again with tremendous *élan*'. The report added that 'fighter tacticians must have been impressed when, during a slow fly-past with wheels and flaps down, Heyworth suddenly retracted everything, opened up the Avons to full thrust and made a near-vertical exit from the arena.'

RA491's 'Avon' career saw it operating mostly with Rolls-Royce, but from May to July 1950 it flew with NGTE at Bitteswell for operations from hard runways (where it joined the Trent Meteor EE227). After 18 months of testing RA491 was replaced by Canberra test beds. Following the 1950 Farnborough it was provisionally allotted instructional airframe number 6879M for use at Cranwell, but instead went from Hucknall to Air Service Training (AST) at Hamble by road, arriving on 1 October 1951. It was then sold to France and, after 5,181lb SNECMA Atar 101.B2 units had been installed by AST

The Rolls-Royce Avon-powered RA491 had nacelles positioned slightly lower than central to permit the wing spars to curve over the top of the engine. The aircraft is pictured during a test flight.

Views showing RA491 during its display at the September 1950 Farnborough Show.

(some sources give the static sea-level thrust rating as 6,000lb), this experimental machine was taken on another 'first flight', from Hamble, on 31 October 1952. Basically, still a Mk.4, RA491 had been modified to Mk.8 'standard' with the latter's forward fuselage and Martin-Baker ejector seat and non-standard wing spars (SNECMA's engineers requested an ejection seat in view of potential lateral control problems from engines providing 33 per cent more thrust than standard Derwents). *Flight* reported that the Atar installation was similar to that for the first Sapphire engines (below) and with the French units in place the all-up weight with a ventral tank was 19,768lb.

On its first flight RA491 was taken from AST to Boscombe Down by Peter Fowler, AST's test pilot, and by mid-November it had completed 3.5 flying hours at Boscombe. Despite the expected problems with lateral stability RA491 was delivered to French engine firm SNECMA at Melun Villaroche on 27 March 1952, French markings and numbers having been substituted for the RAF roundels and registration. It appears that the Meteor had a much shorter operating life here because of an apparent

Comparison views of RA491, firstly with its two Rolls-Royce Avons and second with the SNECMA Atar powerplant. The shape of the nacelles had to be altered to complete the latter installation. The second view was taken at Air Service Training at Hamble.

Side view of RA491 with Atars installed. (Air Service Training)

poor performance, and it completed only eight flights before being grounded in December 1952 and scrapped (French test pilot Marel did indeed report problems with heavy lateral control). Atar engine development would continue with Dassault Ouragan and Mystère fighters, which were not available when RA491 was first earmarked for the role.

Screamer

On 16 June 1950 Mk.8 VZ517 went to Rolls-Royce for development work on the standard Derwent 8 engine fitted in this Meteor mark, initially to help solve problems with surging. Then on 18 July 1953 it joined Armstrong Siddeley Motors at Bitteswell for trials with the firm's controllable-thrust 8,000lb Screamer rocket motor to be used by forthcoming mixed-powerplant fighter aircraft such as the Avro 720 (later cancelled). VZ517 made its first flight with the rocket housed in a re-stressed ventral fuel tank on 17 September 1953 and continued in this role until December 1955 (some sources state that flight clearance was given only in December 1955). After the Screamer rocket project was cancelled in March 1956, in April VZ517 was moved to RAF Halton to serve as ground instructional airframe 7322M.

Right and overleaf above: Armstrong Siddeley photos of VZ517 taken after the installation of the Screamer rocket motor beneath the fuselage. They are dated 22 September 1956. (Phil Butler)

Lift Jet

Standard FR Mk.9 VZ608 was another Meteor used for Derwent afterburner development by Rolls-Royce Hucknall, from 15 March 1951. From June 1953 it was used for reverse-thrust trials and then, having become surplus, in June 1955 it had a Rolls-Royce RB.108 jet-lift unit installed in the main fuel tank bay in the centre fuselage behind the cockpit (the Derwents were still in place). This conversion was undertaken by F.G. Miles Ltd and the objective was to use the Meteor as a flying test bed to develop the RB.108 for its role in the forthcoming Shorts S.C.1 V/STOL research aeroplane. The RB.108 was allowed to swivel +/-30 degrees about the vertical position and when VZ608 was taxying the lift engine's nozzle was just 30in from the ground surface.

After assembly it was test flown from Tangmere before flying to Hucknall. The first test flight with the lift jet in place (but not lit) was made on 18 May 1956 and the trials programme began on 2 August with a short series of flights to develop the air intake for the S.C.1's propulsion engine. The RB.108 was first started in the air on 23 October 1956 and full development of the S.C.1 lift engine began in July 1957. An intake and exhaust arrangement had been developed that permitted the swivelling RB.108 to run satisfactorily under what were quite rigorous conditions. The RB.108 was not usually lit until after take-off, but it was found that the maximum available vertical thrust was sufficient to hold VZ608 at its correct altitude. Early on, there were problems with overheating, which required modifications, and relighting was sometimes difficult. The flying programme embraced 540 tests over 892 lift engine operations in all, the RB.108 accumulating 135 hours and 45 minutes running time.

With all flying completed between 10 May 1960 and 26 June 1964 VZ608 with its RB.108 was used for important ground tests at Hucknall (record cards indicate that this programme actually started in September 1962). The aircraft served as a non-flying test rig to assess both ground erosion and the recirculation of hot gases during, for example, a rolling take-off. VZ608 was taxied over different surfaces at various speeds while using different thrust settings and vector angles. A total of 24 tests and 58 lift engine operations were recorded, the RB.108 accumulating a further 2 hours and 15 minutes running. VZ608 was Struck off Charge on 29 March 1965 and survives at the Newark Air Museum.

Views of Meteor VZ608 in flight with the RB.108 lift unit installed in the centre fuselage. (Roll-Royce Heritage Trust, Derby)

Armstrong Siddeley Sapphire

The most powerful engines to be installed in any Meteor were two 7,600lb Armstrong Siddeley Sapphire Sa.2 axial units in Mk.8 WA820. The Sapphire was originally the Metropolitan-Vickers F.9 and, having gone to Gloster's Moreton Valence works on 17 March 1950, WA820 first flew with its new powerplant on 14 August piloted by Jan Żurakowski (AST was responsible for the conversion). The installation required substantial airframe strengthening to accommodate all of this power, and in particular the wing front spars had to be heavily reinforced because they had to arch over the power units (the engines could not be underslung since this would have required a very long undercarriage).

On 30 August Żurakowski displayed the aircraft at a special public event at London Airport, details of the Sapphire itself having been released from the 'Secret List'. WA820 joined Armstrong Siddeley Motors on 21 March 1951 and appeared with its Sapphires at the 1950 and 1951 Farnborough Displays, Jan Żurakowski flying the aeroplane at the first of these events where he was up against Jim Heyworth in the Avon-powered RA491. In fact, some disappointment was expressed that his performance 'did not exemplify the available thrust of the Sapphires in the manner of Heyworth, that is, by sustained near-vertical climbs from take-off or slow-speed level flight'. Instead, Żurakowski adhered to his planned sequence of manoeuvres, the most spectacular of which was 'a series of three loops from take-off which won acclaim from experienced Meteor pilots'.

Trial flights out of Moreton Valance and Bitteswell confirmed that this was the fastest Meteor to have flown. Armstrong Siddeley's publicity for the 'giant' engine stated that 'the Sapphire has run the 150-hour service test at 7,200lb thrust. This is 1,000lb thrust greater than any jet engine flying today.' To back this up, on 31 August 1951, flying out of Moreton Valence, WA820 established a new time-to-height climb record of 3 minutes 9.5 seconds from a standing start up to 39,370ft (12,000m). In the cockpit was Armstrong Siddeley test pilot Flt Lt R. B. 'Tom' Prickett and a height of 9,843ft (3,000m) had also been reached in 1 minute 15.5 seconds, 19,685ft (6,000m) in 1 minute 50 seconds,

Left: There appears to be an abundance of photographs showing WA820 with its Armstrong Siddeley Sapphire installation. Some of the best are included here and this superb colour image, with both the subject aircraft and the Meteor T 7 (with photographer Russell Adams in the back) performing a loop together, appeared in a Gloster Aircraft poster. (Jet Age Museum)

Below: This view of WA820 in its hanger may have been taken prior to the aircraft's first flight with Sapphire engines. The Hawker Siddeley Group nose logo has not yet been applied to the nose. (Mark Roberts)

Splendid sequence of pictures showing WA820 during a publicity flight in 1950, early in its career. Note the large tail 'bumper', fitted to prevent the rear of the engine nacelles striking the ground during nose-high landings. (Jet Age Museum)

Above: Meteor WA820 pictured at rest. The massive nacelles were as large as portions of the fuselage. (Rolls-Royce photo)

Above: WA820 photographed, it is thought, at the 1951 Society of British Aircraft Constructors (SBAC) Display. The Hawker Siddeley Group logo on the nose shows up well. (Terry Panopalis collection)

Left: The Sapphire Meteor, photographed possibly on 17 August 1951. Test pilot Flt Lt R. B. 'Tom' Prickett, who set a new time-to-height climb record in this aircraft, stands alongside. The size of the nacelles is again most prominent.

and 29,528ft (9,000m) in 2 minutes 27 seconds (when a standard Meteor Mk.8 needed over six minutes to reach 30,000ft). WA820 was used primarily by the engine manufacturer, being based at Armstrong Siddeley's Bitteswell airfield from March 1951 until April 1952. It was also flown at the Central Fighter Establishment, West Raynham, during May and June 1952 to simulate the climbing performance of a rocket-powered interceptor fighter. After these flights the aircraft was returned to Bitteswell in July 1952, and then in late April 1954 retired to RAF Halton as instructional airframe 7141M.

Rolls-Royce Soar

Another Mk.8 employed as an engine trials aircraft was WA982, which had an expendable Rolls-Royce RB.93 Soar jet engine mounted on the port wing tip. Previously it had flown trials at Farnborough and Boscombe Down, before arriving at Hucknall in November 1952 to have this very small engine installed (which was intended primarily for use in target aircraft and missiles). The Soar was attached directly to the wing tip without any cowling, though it could each be protected by a light alloy cover that slid on to the intake. The objective was to access the engine's behaviour up to 45,000ft.

Above: Images of WA982 with Soar units attached to its wingtips also seem to be plentiful. This shot was taken at Hucknall. (Roll-Royce)

Right: Close-up of the Soar installation.

For the 1954 SBAC Farnborough Display WA982 was demonstrated with two Soars and as such became the first four-engined Meteor – the starboard engine was installed just for the show and afterwards removed (for the rest of time a fuel supply system was fitted on that side). After completing its Rolls-Royce service in March 1956 (when the RB.93 was cancelled) WA982 went into store at No 8 Maintenance Unit at the end of June. From 25 March 1957 it spent a short period with NGTE on development work and then on 2 July 1959 was passed to Flight Refuelling Ltd for conversion into a Mk.16 drone.

Note: Another 'four-engine' Meteor was the French night fighter NF.Mk.11 serial 'NF11-3', which in 1954 was modified to take two 1,323lb SFECMAS S-600 ramjet engines mounted in pods outboard of the Derwents. The following year the S-600s were replaced by larger 2,513lb thrust S-900s.

Beautiful views of the Soar Meteor taken for publicity purposes by *Flight*'s cameraman. Particularly well shown is the arrangement of the nose gun armament, which in many other Meteor test beds was either removed or faired over.

Above and below: WA982 on view at the 1954 Farnborough Show. The majority of views of this aircraft, publicity shots for Farnborough or private images taken during the event, show four engines (two Derwents and two Soars), when for most of its career it flew with just three (one Soar). The colour image shows the neat light alloy cover used to protect the intake. (Terry Panopalis collection)

Below: A second 'four-engined' Meteor was an NF11 night fighter acquired by the French, serial 'NF11-3', which in 1954 had a ramjet mounted in an underwing pod outboard of each Derwent.

Chapter 2
De Havilland Vampire

Vampire TG278 pictured on 23 August 1945 before its de Havilland Ghost engine had been installed.

Several de Havilland Vampire fighter aircraft found employment as engine test beds and this chapter reviews the histories of six examples in three separate programmes – the installation of a more powerful de Havilland Ghost engine instead of the standard Goblin, the installation of the Rolls-Royce Nene engine, and adding reheat to the Goblin.

Ghost Record Breaker

Vampire F.Mk.I TG278 was fitted with the Ghost and during the trials this aircraft would go on to break a world time-to-height record. The new engine was almost 3in wider than the normal Goblin, which resulted in a slightly wider nacelle, and it was also 5in longer so the jetpipe had to be extended slightly. The Ghost itself was aimed primarily at the civil aircraft market (in particular de Havilland's own Comet) and so its altitude performance in particular had to be assessed. To cope with the difficult conditions in the rarefied atmosphere experienced at great heights TG278 had its pressurised cockpit strengthened by a new metal canopy with side 'portholes' – the cockpit was pressurised from the main engine compressor to give the pilot an equivalent height of 37,000ft when flying at 60,000ft. In addition, to lower the wing loading, its wingspan was increased to 48ft 0in up from the normal 40ft 0in, the new sections having more pointed tips. TG278 made its maiden flight as a Ghost test bed on 8 May 1947.

During a flight made on 31 January 1948 de Havilland chief test pilot John Cunningham reached almost 56,000ft above sea level, and within 61ft of the long-standing world record set by an Italian on 22 October 1938, Colonel Mario Pezzi flying in a Caproni 161. Although the Comet's Ghost engine was to have a static thrust rating of 5,000lb, the development unit installed in TG278 on this flight was rated at just 4,400lb. Then on 23 March 1948, flying out of Hatfield, Cunningham recorded a World

Above: The experimental high-altitude Ghost Vampire TG278 after a metal canopy had been added. The aircraft's nacelle was now slightly longer and the rear nacelle section a little bulkier. (Phil Butler)

Below: The extended wing tips fitted to TG278 for its altitude record flight. (Phil Butler)

Below: TG278 on display at the 1948 Farnborough Show, having been painted white.

Record 59,446ft in the Ghost Vampire. The de Havilland team had calculated that each pound saved in weight would be worth 2ft in altitude and so the additional measures taken to cut weight included the removal of the aircraft's guns, the radio and all of the exterior paint. The take-off weight, with a total of 202gal of fuel aboard, came to 8,400lb and this capacity was sufficient for an hour's flying. TG276 was capable of a maximum rate of climb of about 8,000ft/min but, in order to avoid the risk of lag on the critical instrument readings for the record-breaking sortie, a figure somewhere between 3,500ft/min and 4,000ft/min was selected. The record flight lasted 47 minutes.

Looking at TG278's career as a whole, back in July 1945 it had been allotted to de Havilland for a trial camera installation as a prototype photo-reconnaissance version of the Vampire. In October 1945 it was earmarked for conversion as the Ghost II flying test bed at Hatfield, but it was 1947 before this became reality. In early September 1948 TG278 appeared as a static exhibit at the Farnborough Show newly repainted in an all-white scheme. Then on 2 October 1950 the aircraft suffered an 'accident at Hatfield', and on 6 March 1951 it arrived at No 1 School of Technical Training at RAF Halton to serve as ground instruction airframe 6851M.

Nene Trials

Four British Vampires, serials TG276, TG280, TX807 and VV568, had a 5,000lb thrust Nene installed, a far more powerful unit than the Goblin. Installing a Nene to produce what was to be called the Vampire F.Mk.II (later F.2) should have provided an improved performance, but there was the problem of getting good airflow from the air intakes to the engine. From a size point of view the Nene was similar to the Goblin, so installing it in the fuselage was reasonably easy (the aircraft's span and length were unchanged). It was the nature of the design of the Nene itself that brought the difficulties because this had a double-sided impeller and that required air to be routed both to the front and to the rear of the engine, in contrast to the Goblin with its single-sided impeller that could be served purely by the wing root intakes. Overall, the Nene Vampire's performance proved to be down on estimate by some margin and modifications to the intakes did not help.

The standard Goblin intakes did not supply enough air for the Nene and so trials followed with supplementary 'elephant ears' intakes introduced behind the cockpit. These helped but increased the drag, while a flight test report from October 1949 revealed that the modification had also resulted in some pretty unpleasant handling characteristics, especially so at high Mach numbers, which made the aircraft unacceptable. Compressibility (early indications for approaching the speed of sound) at

This well-known in-flight view of TG276 with a Rolls-Royce Nene installed shows to advantage the supplementary 'elephant ears' air intakes, made necessary by the engine's double-sided impeller. (Rolls-Royce)

TG276 as a Nene Vampire Mk.II, taken at Samlesbury in July 1946 before the 'elephant ears' had been added. Note the high tailplane.

high altitude appeared at a lower Mach number than for a standard aircraft without the 'ears' and the resulting vibration experienced above Mach 0.75 at 15,000ft, together with high-frequency elevator and aileron buffet and a lack of stability overall, was considered unacceptable (although it was not violent). The report added that 'below this altitude at the high indicated airspeed range, limiting Mach number is determined by a different vice, namely a severe nose down trim change which builds up very rapidly'. In general, Vampire II's handling at both high and low indicated airspeeds and with compressibility present was 'most unsatisfactory' and it appeared that the cause was the 'elephant ear' cowlings. These characteristics were most severe when the Nene was throttled back or if the engine rpm was dropping, and this particular condition also produced elevator buzz and an associated spongy feel. The original intakes with the Nene had given a maximum speed of 513mph at 5,000ft and 515mph at 25,000ft, the 'elephant ears' installation increased these figures to 533mph and 519mph respectively, but in present form they were clearly unsatisfactory. As a result, Rolls-Royce fitted flaps inside the 'elephant ears' and in November 1949 Cunningham tested the aircraft with the 'elephant ears' blanked off, i.e. to simulate the flap shut condition. He found that the blanking off removed the more serious vibration when the

Further air-to-air views that offer detail of the aircraft's 'elephant ears' intakes from different angles. (National Aerospace Library)

aircraft was throttled back at high Mach numbers, as well as the buzz and sponginess when flying at low speeds and altitudes. However, a sharp nose-up jerk still occurred when throttling back at high indicated airspeeds and there was still a serious nose-down trim change at Mach 0.73 when flying below 10,000ft. The high frequency buffet or buzz also appeared at high altitudes upon reaching Mach 0.74 with power on.

In the end the F.2 was never acquired by the RAF, but the Royal Australian Air Force (RAAF) did order Vampires fitted with the Nene instead of the Goblin as the F.Mk.30 and, after looking at larger air intakes, it returned to 'elephant ear' intakes fitted with spring-loaded flaps. France also acquired Nene-powered Vampires but these used enlarged wing root intakes based on those of the Hawker Sea Hawk naval fighter; the design work for this modification was undertaken by Boulton Paul and proved very successful. Originally, with a Nene and 'ear' intakes, and at a take-off weight of 10,700lb, the top speed was 532mph at sea level and 486mph at 40,000ft, with time to that height 12.2 minutes; with the Hawker-type intakes these figures became 10,900lb, 566mph, 489mph and 9.2 minutes.

Moving to the individual careers, TG276 was allotted to Rolls-Royce for installation and flight development with the Nene in June 1945 and flew into Hucknall by air under Goblin power on the 27th. It had initially been intended to convert the third Vampire prototype, MP838, but then it was decided to use a production machine. The first flight under Nene power came in March 1946 and,

This picture of TG276 was taken at Boscombe Down in February or March 1950 after the aircraft had received larger Hawker Sea Hawk jet fighter-style wing root air intakes to feed the Nene.

from this point until well into 1948, the aircraft spent time both at Hucknall and at A&AEE Boscombe Down, suffering damage on two occasions including catching fire in the air on 14 December 1946. On 11 February 1948 TG276 had to force-land at Boscombe Down due to engine failure, and on 12 November it was allotted to Boulton Paul for the air intake modifications, being moved from Hucknall to Wolverhampton by road on 10 December. It returned to Boulton Paul again on 4 February 1950 to have a new low tail unit fitted, and on 31 March 1950 TG276 was allotted to the French government, being despatched to SNCAN (Nord Aviation) at Meaulte on 5 April. It was finally sold to the French government on 30 December 1953.

TG280 was also allotted to Rolls-Royce for Nene installation and flight development in June 1945, arriving at Hucknall from Preston on 22 September. Initially, however, it flew with its Goblin to provide comparison data with the Nene powerplant. It was despatched to de Havilland at Hatfield for Nene flight tests on 6 July 1946, but it was extensively damaged in a ground accident at Hatfield on

Nene Vampire TX807 pictured at Samlesbury on 9 January 1947 without 'elephant ears'. (Crown Copyright)

Left: When this beautiful photo was made TX807 had received its additional intakes. (Phil Butler)

Below: The last British Nene Vampire VV568 is seen here in October 1949 at Hatfield (identified by the hexagonal pattern of the concrete sections). This aircraft had 'elephant ears' intakes and, as a Mk.5 Vampire, it also had the low-position tailplane.

17 October. After repair at Hucknall, in 1947 TG280 flew development trials of 'revised air intakes for high Mach numbers', both with Rolls-Royce and A&AEE Boscombe Down. Next, on 20 April 1948 it went to English Electric at Samlesbury to have a modified tailplane fitted, and then on 20 July to Hatfield for investigations into stalling characteristics. On 14 February 1950 TG280 made a wheels-up landing at Hucknall, which brought Nene Vampire testing by Rolls-Royce to a close. Finally, on 15 June 1951 this Vampire was despatched to RAF Cranwell, minus its engine, to serve as Ground Instructional Airframe 6797M.

In May 1947 TX807 was allotted to Rolls-Royce to test 'additional intakes to improve engine efficiency', and then in October it went to Boscombe Down for handling checks prior to despatch to Australia. On 12 August 1948 it arrived in Australia aboard the SS *Northumberland* and was subsequently received by No 1 Aircraft Depot (AD) at Laverton, and then on 2 September it was taken on charge by the RAAF (with serial A78-2) and allotted to the Service's Aircraft Research and Development Unit (ARDU), also based at Laverton. A78-2 arrived in Australia while production of the local Nene-powered Vampire F.30 was underway. The ex-British airframe seems to have spent most of its time with ARDU, but also at least a short period with No 78 Wing based at Laverton, which was the parent unit for the Vampire squadrons (note: UK-built Vampires had A78-numbers, Australian-built A79-numbers). From October 1949 de Havilland Australia fitted A78-2 with new wings with larger air intakes, which enabled the dorsal intakes to be removed. Further time was spent with 78 Wing and ARDU before, on 10 August 1956, the aircraft was grounded on the completion of its test programme. It was then allotted to ground instructional duties at RAAF Rathmines as 'Instructional No 2' and in December was transferred to the RAAF School of Technical Training at Wagga Wagga. In September 1961 it was allocated for firefighting training.

The last of the Nene test beds, VV568, was used in the development of France's Nene Vampires, undergoing conversion at Hatfield to French standard as a trial installation from February 1949 onwards. On 14 December 1949 it arrived at A&AEE from Hatfield for handling and stability checks prior to handover to France and was then despatched 'airframe only' to its new owners on 21 January 1950.

Goblin Reheat

In May 1949 Vampire FB.Mk.5 VV454 was fitted with an experimental Goblin 2 engine equipped with reheat in an extended rear fuselage. This aircraft had the Mk.5 wing, but it used the old square-cut fin and rudder assemblies and high tailplane of the Mk.1 to keep the horizontal surface clear of hot gases; special skids were added beneath the booms to guard against the long tailpipe impinging on the runway. Already mentioned in Chapter One, exhaust reheat (and known in America as afterburning) was the name given to the relatively new process of burning fuel in the exhaust pipe of a jet engine as a means of increasing the exit gas velocity and hence its thrust. By increasing the final nozzle area it was possible to burn fuel in the jet pipe without imposing any additional stresses on the turbine assembly. Before its flight trials began VV454 was tested on a special ground rig. It was also shown at the September 1949 Farnborough Show where it was displayed by Chris Beaumont, chief test pilot for de Havilland Engines. The *Flight* magazine report of this display highlighted in particular the impact of the tremendous noise created by reheat and noted that the 'Vampire adaptation represents an early step in de Havilland's reheat development programme, but Mr. Beaumont showed that this step is an appreciable one. When reheating sets in, an appalling din blasts the ear drums, but quickly diminishes as the Vampire recedes aloft'.

Previously, from mid-November 1948, VV454 had been on the strength of the Central Fighter Establishment (CFE) at West Raynham. In March 1949 de Havilland began using this Vampire in an investigation of jet engine fuel systems using petrol (rather than kerosene/paraffin) and then in June it was officially allocated to the reheat programme. After the latter project was completed VV454 was restored to its normal FB.5 configuration, and finally in June 1953 it was passed to de Havilland at Hatfield for instructional use.

The reheat Goblin Vampire VV454.

Above: VV454 featured the high tailplane used by the Vampire Mk.I.

Above and below: The reheat Vampire at the Farnborough Airshow in 1949. (National Aerospace Library)

VV454 on view at Hatfield on 5 September 1949, showing clearly the skids attached to the undersides of the booms.

This near end-on view of VV454 provides good detail for the reheated engine's jetpipe. (National Aerospace Library)

A poor quality but very rare photo showing VV454 with the reheat lit. The value of the high position tailplane, well above and out of the way if the very hot exhaust, is clear. (National Aerospace Library)

Above left: VV454 receives maintenance at Hatfield on 24 August 1949. (de Havilland Engines)

Above right: This close-up of VV454, taken at Hatfield on 5 September 1949, provides excellent detail of the Goblin jetpipe and the wing flap arrangement. (de Havilland Engines)

An interesting picture of VV454 on display with other de Havilland products – the Vampire night fighter prototype G-5-2, another Vampire and, at the back, a Comet airliner, in this case G-ALVG.

Chapter 3
Hawker Aircraft

Hawker P.1072

The 1950s witnessed a brief but extraordinary episode in fighter development when the UK's aircraft industry tried to put a rocket-powered fighter and interceptor into RAF and Royal Navy service. This effort resulted in the Saunders-Roe SR.53 research aircraft of 1957, but groundwork had been put in some years earlier by the Hawker P.1072.

The first of three Hawker Aircraft P.1040 jet fighter prototypes, VP401, first flew on 2 September 1947 and the design was subsequently turned into the Royal Navy's successful Sea Hawk. After completing its development flying the now-redundant VP401 was adapted to enable it to operate as a combined rocket and jet-powered research aeroplane. As a proven airframe and powered by a proven jet engine, the P.1040 made an ideal test bed for trials with an additional rocket motor and in June 1949 VP401 was allotted for conversion. It was returned to Hawker's Langley facility for a quite extensive series of modifications to fit an Armstrong Siddeley Snarler rocket motor to compliment the P.1040's existing Nene jet. In this form VP401 received a new Hawker project number, P.1072. Development of the 2,000lb thrust Snarler had begun in 1947 and the motor was described as a 'hot' rocket because it ran on a water/methanol mixture (35 per cent water and 65 per cent methyl alcohol by weight) with liquid oxygen as the oxidant. The unit was designed to provide a hefty increase in an interceptor-type aircraft's rate of climb and (for short periods only) to improve its performance at very high altitude (where the rocket would maintain full power). The Snarler could run for a period of 2.75 minutes duration.

The conversion meant that the kerosene fuel for the 5,000lb Nene RN.2 had to be reduced by nearly half to make room for the rocket motor fuel. There was now 76gal in the forward tank, 39gal in a saddle tank and another 60gal in the rear tank. A total of 75gal of liquid oxygen was contained in a

The Hawker P.1072 prototype VP401 after it had been painted in duck-egg green livery.

cylindrical tank placed behind the pilot and forward of the engine bay, with 120gal of methanol water in the aft part of the rear tank bay. The new main supply piping from the pumps to the motor itself went inside an external spine running along the bottom fuselage centreline and installing the rocket pumps required some rearrangement to the Nene pump and accessory bay. The methanol feed pipes ran forward to the pumps ahead of the Nene and then back to the rocket in the tail, and the rocket exhausted at the extreme end of the fuselage. Much of the new piping was made in stainless steel.

VP401 in its original form, as a P.1040 prototype, flying at the Farnborough Show in 1948.

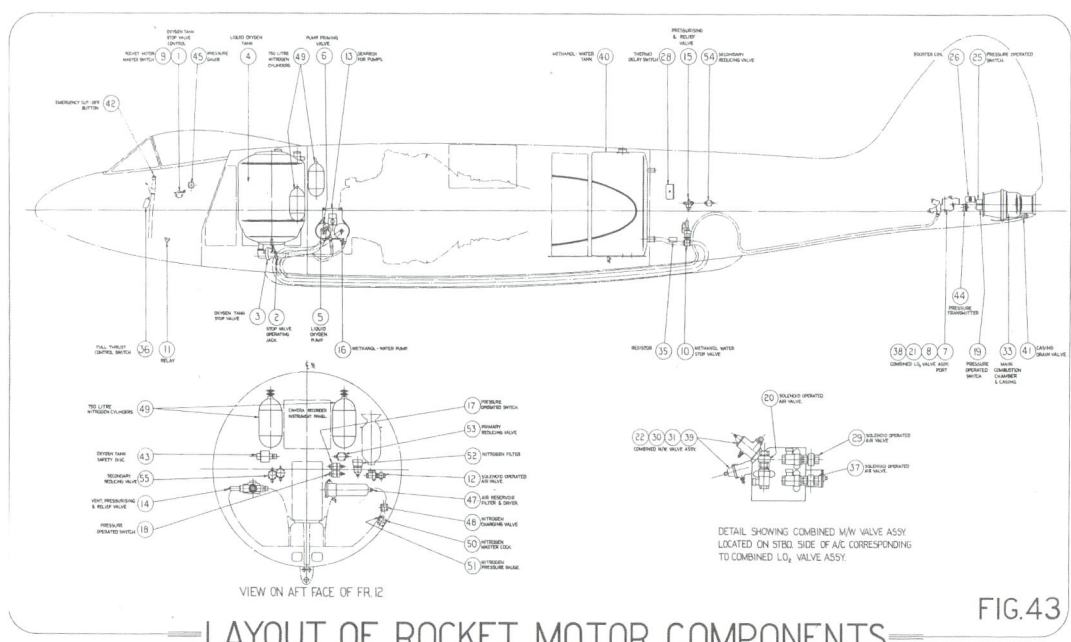

Drawing showing the layout of the P.1072's rocket motor components. (Chris Farara)

Hawker's Reg Smyth designed the new installation that weighed approximately 213lb, with the aircraft's maximum weight now coming to 14,050lb. Extra ballast had to go in the aircraft's nose to balance the weight of the rocket motor and its associated fittings in the rear and VP401 was also fitted with an adjustable tailplane. The fin area was increased and there was now a bullet fairing at the fin/tailplane leading edge junction, a feature that VP401 did not have as built but which eventually appeared on the Sea Hawk. It was decided to use the original wings but to reinforce them to cater for the increased wing loading of the conversion. However, the effort to develop and clear the ASSn.1 rocket motor in VP401 for flight took longer than expected. Indeed, there were periods when VP401 stood idle in the Experimental Shop awaiting completion and its motor did not arrive from Armstrong Siddeley until June 1950. When completed the aircraft had an all-metal finish, but later it was painted in the duck-egg green used by Hawker for several prototypes during the 1950s.

Right: This view of VP401 provides detail of the rocket installation in the end fuselage and the faired duct or keel introduced along the fuselage underside. (Hawker Siddeley via Terry Panopalis)

Below: Side view of the P.1072 as first converted. (Hawker via Terry Panopalis)

Above left: VP401 is refuelled, or is undergoing tests, on 30 August 1951, for which it has the under-fuselage piping uncovered. (Hawker)

Above right: The P.1072's rocket motor is fired during ground testing. (Hawker)

VP401 first flew in the new configuration on 16 November 1950, but this sortie was a ferry trip to the Armstrong Siddeley Motors' flight test facility at Bitteswell using the Nene alone. After some further ground runs, the first flight with the Snarler fired during the sortie came on 20 November in the hands of Hawker chief test pilot Sqn Ldr Trevor Wade. In 1954 *Flight* magazine reported how 'the initial flight of the P.1072 was the first ascent by any pump-fed liquid-oxygen rocket in a piloted aircraft. The day was dull and overcast, and the P.1072 could be heard only distantly through dense cloud. Those on Bitteswell airfield were quite ignorant of whether Wade had fired the rocket or not and were still waiting expectantly when the fighter appeared low down at full power. Passing overhead, Wade did a smart roll; all had gone well.' Hawker's own diary reported 'all rocket fuel used with satisfactory results.' Five more flights were completed from Bitteswell up to 19 January 1951 with generally very satisfactory results. However, during the last of these a faulty pressure gauge transmitter exploded, which started a minor fire in the tail assembly. The pilot on this sortie was Hawker's Neville Duke, who shut down the Snarler before making an emergency landing at Bitteswell on just Nene power. The rear fuselage had

The spectacular sight of VP401 taking off from Bitteswell for a test flight with the rocket motor running. (Hawker)

Top: Refuelling the P.1072 prototype at Bitteswell in 1951.

Above: Here the refuelling operation is being conducted with the methanol feed pipes uncovered. (BAE SYSTEMS Heritage)

Right: The sole P.1072 prototype pictured on static display at the 1951 SBAC Farnborough Show. (Terry Panopalis)

suffered appreciable minor damage, but by mid-February this had been repaired and the first ground run following this incident was made at the end of the month.

In the air the rocket was not throttleable and in addition the cockpit was unpressurised, which restricted the flying to altitudes below about 30,000ft. The time required for the Snarler motor to reach its maximum thrust would vary with different aircraft installations, but for the P.1072 it took around eight seconds. The airframe's critical Mach number also meant that the rocket could normally be fired only during a climb but, with the reduced thrust of the Nene at 30,000ft, firing the Snarler did increase VP401's rate of climb by as much as five times. The rocket was also fully aerobatic.

On 19 February 1951 VP401 was handed over to Armstrong Siddeley for its official Snarler trials. On 22 March it was transferred to Directorate of Engine Research and Development (DERD) control, and on 16 April 1952 it arrived at RAE Farnborough from Bitteswell for temporary storage. In August 1954 *Flight* reported that 'experience with the P.1072, and other developments, has now taught Armstrong Siddeley a great deal about rockets for piloted fighters. It has been found, for example, that an aircraft with such a motor can be held at readiness with the tanks filled and that – if they are carefully designed – the valves do not freeze up.'

Valuable knowledge in the use of rocket motors had been provided by the P.1072, but official interest in the Snarler gradually waned and this rocket motor project was subsequently abandoned. VP401 did appear in the static park at the 1951 SBAC Farnborough Show but, quoting *Flight* again, 'to everyone's regret the P.1072 remained silent and static at the Show.' From 8 August 1952 the sole P.1072 was used for 'minimum drag' flight tests, on 10 August 1954 it was Struck off Charge as 'weight of metal', and in the autumn of that year the aircraft was scrapped. In its P.1072 form VP401 can be termed a success, though the Snarler motor gave too little thrust to be really of much value. P.1072's span was 36ft 6in, length 37ft 7in and gross wing area 256sq.ft. In the air it had a maximum level speed of Mach 0.85 at 36,000ft, it took 10 minutes 30 seconds to reach 35,000ft, and the ceiling was approximately 45,000ft.

Hunters

The second production Hawker Hunter F.Mk.5, serial WN955, was employed as a test bed for Armstrong Siddeley's Sapphire Sa.7 (200-series) engine as (designated by *Flight* magazine) a 'Mk.6 standby', presumably against failure of the Rolls-Royce Avon 200-series for the Hunter F.Mk.6. The Sa.7 was placed in the same class as the Avon RA.24, it would fit readily into the Hunter fuselage and with a thrust rating of 11,000lb was expected to give the same performance (although changes to the air intake would be required). The Mk.5 was normally powered by a Sapphire Sa.6. Having flown for the first time on 26 October 1954, WN955 joined the Ministry of Supply's test fleet at Bitteswell on 9 November and during 1955 was converted to receive the more powerful Sapphire, making its maiden flight with this in place on 9 February 1956 with Armstrong Whitworth pilot Flt Lt W. H. Bill Else at the controls. In this form it was painted silver, carried Class B mark 'G-1-2' and was employed on Sapphire 7 trials until 8 March 1956. But its research career was far from over – indeed many combat aircraft like this never joined a squadron but instead spent their entire working life on experimental work. (Note: there is no known photo of this Hunter flying as a Sapphire 7 test bed. Also, a list of Bitteswell flying test beds states that WN955 was used for 'Sapphire reheat development, in conjunction with a Private Venture Armstrong Whitworth Aviation project for a modified Hunter'.)

From November 1956 A&AEE used WN955 for banner and sleeve target towing, almost certainly with the standard Sa.6 restored. On 16 April 1958 it went to Farnborough to investigate the feasibility of air-launching banner targets from special containers mounted on the wing bomb pylons and towed from units positioned centrally under the fuselage. Then on 16 July 1959 the Hunter joined RAE Farnborough's Radio Department for trials, with the aircraft running its engine on the ground as part of some infra-red experiments. Struck off Charge on 15 July 1960, WN955 was destroyed on the Farnborough fire dump in 1967.

The Hawker P.1099 (F.Mk.6) prototype, serial XF833, made its maiden flight on 22 January 1954 and went to A&AEE for handling trials on 20 April. From 7 June 1955 it spent a period at RAF Wymeswold for trials with its Avon RA.28 engine, before going to Boscombe Down towards the end of July. Next it was used in a Ministry of Supply development programme to assess a Rolls-Royce thrust reverser, because the Hunter's centre-fuselage jet pipe made it perfect for the trial. To this end XF833 was provided with an orifice on each side of the jet pipe in which was mounted a pair of hemispherical eyelid shutters. When reverse thrust was applied these would close and all of the RA.28 engine efflux was then forced to exit through new lateral rectangular outlet grills in the sides of the rear fuselage, each of which had multiple cascade vanes positioned such that they would deflect the exhaust outwards and forwards.

The conversion, which involved minimal external modification to engine or airframe, was carried out by Miles Aircraft at Shoreham from June 1956, the Hunter having arrived on the 9th of that month. In July XF833 was delivered by road to Hucknall, on 26 August it made its first flight with the reverser operating in order to reduce the landing speed, and seven such flights were completed before the aircraft went to the September 1956 SBAC Show. Here a smoke generator bottle in the tailpipe was used to show the direction of gas flow (which at times completely immersed the aircraft in smoke). In the cockpit for the show's demonstration flying was Rolls-Royce chief test pilot A. J. 'Jim' Heyworth.

The initial landing tests were performed during November and December 1956 and involved almost 130 reversals and, since Hucknall's runway length did not permit no-brake landings, all of these were made using full braking plus the reverser. Furthermore, a speed limit was imposed because the reverser could not be operated at speeds of less than 80 knots due to control surface flutter and aileron buffeting, caused by the forward thrust of the exhaust efflux. The landing roll distance – using full braking without reverse thrust – came to 2,520ft, while the average figure with the reverser operating saw this reduced by 919ft. Altogether 308 landings were completed before XF833 went to RAE Bedford on 21 April 1958. The airframe was then taken to Farnborough on 17 November 1958, it was Struck off Charge on 31 October 1962 and apparently scrapped at Farnborough in 1963. The programme's objective had been to prove the operation and reliability of the thrust reverser, with its principal applications connected to the Rolls-Royce Conway engine in future long-range airliners.

The thrust-reverser Hawker Hunter test bed XF833. (Peter Green)

Spectacular photo of XF833 with the thrust reverser and smoke generator (to show the direction of gas flow) in operation. (Rolls-Royce Heritage Trust)

Close-ups showing the external appearance of the Avon thrust reverser installation.

Even with a thrust reverser fitted the Hunter still looks beautiful in the air. These photos were taken for publicity purposes on the run up to the 1956 Farnborough Airshow.

Above: **At one stage XF833 had two black bands painted around the fuselage for photo-interpretation reasons (the second just behind the wing-trailing edge). (Peter Green)**

Above: **XF833 comes in to land at Farnborough in 1956.**

Below: **The test bed Hunter on the runway at Farnborough in 1956 with flaps deployed and just at the point where the thrust reverser would be activated.**

Chapter 4
Gloster Javelin

Gloster Aircraft photograph showing Javelin FAW.1 XA552 in its original form powered by Sapphire engines. This was taken when the aircraft was performing trials for the parent company. (Jet Age Museum)

Gloster Aircraft's final production jet aircraft was the Javelin transonic night fighter and two FAW.Mk.1s were adapted to take non-standard engines.

De Havilland Gyron Junior

The Gloster Javelin Mks.1 to 6 were all powered by two Armstrong Siddeley Sapphire Sa.6s, the Mk.7 had a more powerful version of the Sapphire and the FAW.Mk.8 introduced afterburning. It is understood that there was never a strong possibility of production machines receiving a different make of powerplant but an early production Mk.1, XA552, flew trials with the de Havilland Gyron Junior, an engine which with afterburning had been selected to power the Bristol 188 supersonic research aircraft.

Earlier, consideration had been given to fitting a Vickers Supermarine Attacker or Swift fighter with a Gyron Junior P.S.50 to act as a flying test bed for both jet deflection and reheat. Vickers made a rapid appraisal of the problems involved with such an installation and, when de Havilland Engines staff visited the firm at Hursley Park on 21 December 1955, they saw a rough layout showing the alterations necessary for an Attacker fuselage to accommodate jet deflection. Briefly, it appeared that in both Swift and Attacker a new fuselage would be required, as the engine had to be moved very considerably forward in order to get the reaction from the jet deflection to pass anywhere near the aircraft's CofG. In the case of the Swift this would be in the order of 4ft to 5ft and in the case of the Attacker 7ft to 8ft.

In addition, the rear fuselage would have to be increased in cross section considerably to accommodate the proposed 36in internal diameter reheat pipe. And a new tail unit would be required to counter the moment imposed by the lengthened fuselage, along with a strengthened wing root rib to accommodate the additional loads imposed on it due to the front spar having to be bent to accommodate the deflected jet ('moment' is an expression covering the product of a distance and a physical quantity and accounts for how the physical quantity is located or arranged). Vickers felt that to do such a conversion on either of these aircraft would involve considerable design time, and it would be extremely unlikely if a modified aircraft could be available for flight in less than 18 months to two years. In view of the foregoing, the Ministry felt that the use of either of these aircraft as a flying teat bed for jet deflection was ruled out, but the Swift could be considered for reheat only. However, the Swift suffered design problems of its own in RAF service, so when the time came to find a Gyron Junior test bed the choice was the Javelin.

In preparation for the Gyron Junior to go in the Bristol 188, the area of development in which de Havilland Engines had least experience of high temperature afterburning was in flight, hence the need for a test bed. XA552 never actually joined an RAF squadron, and it was first employed by Gloster Aircraft, for drop tank tests, at Moreton Valance, and these were still ongoing in the first months of 1956. On 11 December 1956 it was allotted to Napier at Luton for the installation of two Gyron Junior DGJ.I0R engines, which at the time were described by the press as 'the first British aircraft powerplants designed for flight at Mach numbers greater than 2.5'. It was despatched to the Napier Flight Test Department at Luton for conversion on 18 December, and then transferred to de Havilland Engines' charge on 5 August 1959. Hatfield would become its base for much of the flying programme.

The Gyron Junior's exceptional thrust/weight ratio, combined with a relatively small frontal area, had (in theory) made the engine a natural choice to power the forthcoming all-steel Bristol 188, and the PS.50/DGJ.10R version featured zero stage, variable stators and steel construction (test flying of the 188 from 1962 would in fact show that the Gyron Junior was not the ideal). De Havilland had also developed a high-augmentation afterburner that would operate at a combustion temperature of up to 2,000K (3,140°F) using a fully variable supersonic nozzle, and the engine's sea-level static rating was 10,000lb dry and with maximum reheat 14,000lb, rather more that the Sa.6's 8,300lb. Calculations had shown that in flight at 36,000ft, at a speed in excess of Mach 2.5, the complete powerplant could deliver a thrust of more than 20,000lb. The two engines fitted in XA552 were, however, low-powered, afterburning versions of the 188 units, the objective of the Javelin trials being to prove the low-speed handling of the Gyron Juniors in readiness for the supersonic aircraft's early flights. The problems of installing these very advanced power units proved to be formidable, with a great deal of time expended in perfecting the control systems to the engine, afterburner and nozzle. Very extensive airframe and systems modifications were needed because the Gyron Juniors were not merely engines but a complete propulsion system, with a set of complicated controls that were designed to match up all of the variables expected within a wide range of flight conditions.

The first ground runs were made on 26 April 1960, but by 4 May preliminary running had revealed that the Gyron Juniors were very sensitive to surge, even with the nozzles fully open from the start. By 11 May further ground runs had confirmed that these engines were impossible to operate without surge, but it was established that this would be expected due to choking conditions at the intake entry, where the diameter was 22.8in. Model tests carried out on Javelin intakes back in 1956 or 1957 had been made using intakes with an entry diameter of 24.8in, and in the light of this a pair of the original type of intake entry, of exactly 24.8in diameter, were obtained from Bedford. One of these was tried for the starboard engine under ground running conditions and the engine operated satisfactorily. By 3 August tests had also been made with a new 26in-diameter intake duct, but XA552 was still not ready to fly and would miss taking part in that year's SBAC Show at Farnborough.

Above: This de Havilland Engines photo of XA552 was taken in early February 1961 and shows it during one of its very early flights (possibly the first) with Gyron Juniors installed. The aircraft has not yet been painted blue. (Peter Green)

Right and below: Rear-angle pictures of XA552 taken when the aircraft was still in camouflage. Note the considerable amount of test and monitoring equipment on view in the second shot. (Rolls-Royce Heritage Trust, Bristol)

Left: Ground-based engine runs at Hatfield with the Gyron Juniors in reheat. Externally, the new units appear to have provided a neat installation. (Rolls-Royce Heritage Trust, Bristol)

Below and bottom: Lovely colour photos of XA552 newly repainted in a royal blue colour scheme with titles on its nose in red. For its test bed role the aircraft had been given a new nose with a long instrumentation boom. (Rolls-Royce Heritage Trust, Bristol)

XA552 goes through preparations for another test flight.

In fact, ground running continued right through into early September 1960, illustrating the difficulty in getting this new installation to work smoothly. By 2 November both engines had been ground run to full duty without surge and by 9 November the runs had been completed with and without reheat. At last, the maiden flight was recorded on 31 January 1961 with John M. Nicholson, de Havilland Engines chief test pilot, in the cockpit, but after 25 minutes XA552 had to put down at Bedford due to a lack of throttle response during dummy approach runs. Later, while on the ground, the system reset itself and the aircraft was then flown back to Hatfield. Four flights had been completed by 6 February, the aircraft handling well up to 20,000ft and accelerating to full speed in 5–6 seconds. Further flights were made during the following week, and an engine relight was achieved at 5,000ft. Flame out had been experienced with one engine on slamming the throttle shut, but on 27 February the reheat was lit satisfactorily on the port engine at 3,500ft and 230 knots.

On 6 April 1961 Godfrey Auty, the Bristol test pilot who would perform the flight trials of his firm's 188 research aircraft, made his first flight in XA552. At 10,000ft he lit the port reheat at a speed of 200 knots, the reheat being operated up to approximately the maximum value. During the next week Auty made seven more flights, including a trip to Bedford where he tested the reheat on the starboard side at 5,000ft and 200 knots. At Bedford two flights were made employing reheat on take-off while, back at Hatfield, on 20 April Auty made a reheat take-off and climb to 10,000ft with full reheat on both Gyron Juniors, the flight including light up and reheat operation to full power at 10,000ft and 300 knots and then 1,000ft and 300 knots. Later the aircraft's general engine handling at 15,000ft and 20,000ft was explored and successful relights were made at 15,000ft at both 200 knots and 250 knots.

From mid-May 1961 XA552 underwent major overhaul and the engines were not reinstalled until mid-August. On 4 September XA552 was taken on shakedown tests so that it could display at Farnborough. By now it had been painted in an attractive dark blue scheme and had completed some 21 hours of development flying. XA552 performed all week at the SBAC Show at Farnborough, being flown everyday bar one by pilot Peter Barlow. The aircraft was back at Hatfield on 11 September and a week later an intensive reheat flying programme was being planned for the PS.50 engine, all preparatory for the Bristol 188's flight test programme. In the week ending 2 October four flights were made covering engine handling and reheat operation between 5,000ft and 15,000ft, all by Bristol test pilot J. 'Willie' Williamson. However, the engines had to be taken out during November due to troubles with seal deterioration and returned to de Havilland Stag Lane. During its test bed career XA552 would spend a lot of time on the ground, and on this occasion the engines were not reinstalled until mid-February 1962.

The next flight was made on 7 March with six more that month, but further flights during April showed that a large quantity of oil was being used. Then, during a ground run on 16 April, a second stage turbine failure occurred in the starboard engine. Both Gyron Juniors were removed and were not reinstalled until 1 February 1963. In the meantime, in June 1962 XA552 had been allotted to Bristol Siddeley Engines at Filton for the trials to continue there, de Havilland Engines having now been taken over by Bristol Siddeley.

This photograph appears to have been taken in 1962 and provides underside paint scheme detail.

The Gyron Junior Javelin is lined up for take-off on the runway at the 1961 Farnborough Show. (Adrian Balch)

Top: Take-off in full reheat at Farnborough, another spectacular sight!

Above: XA552 comes into land at Farnborough after one of its 1961 displays. Note the Lightnings in the background.

Right: In this photo XA552 carries standard Javelin 'bosom' external fuel tanks under its fuselage, which have also been repainted royal blue.

Publicity pictures of XA552's starboard side. (Rolls-Royce Heritage Trust, Bristol)

It flew again on 6 March, to Patchway (Bristol), and ten flights had been completed by the week ending 3 May, after which XA552 was grounded for the repair of a suspected crack in the fabricated instrument ring that was positioned upstream of the engine intake casting. This proved to be the end for XA552's flying career because the weekly report for 10 May 1963 noted 'it is understood that this aircraft has been taken out of the PS.50 development programme and the engines are to be removed and cannibalised for Bristol 188 use'. Indeed, further work had been cancelled on 27 April 1963, the programme being described as 'complete'. By 24 May the aircraft was being cannibalised with the port reheat tailpipe and nozzle already removed for use on the test bed at Hatfield. On 5 July it was reported that XA552 was for disposal with the engines now at Stag Lane, and on 5 December 1963 the Javelin was sold to R. J. Coley for scrap.

It appears that the flying programme proceeded with relatively few troubles and clearly much was learnt about the behaviour of the de Havilland engine. XA552's Gyron Juniors did suffer from surging, as did the units flown later in the Bristol 188, but the reports covering nearly 60 flights known to have been logged by XA552 as an engine test bed confirm that there were no major accidents or incidents or critical problems at all. There appears to be some uncertainty as to just how XA552 sounded when

XA552 photographed in flight from the port side, almost certainly during the same sortie as the other air-to-air views. (Rolls-Royce Heritage Trust, Bristol)

it was flying, the aircraft apparently making an altogether different noise to the standard Javelin. *Flight* magazine, when reporting on the 1961 Farnborough Show in its 7 September issue, described a whine or howl, while one witness has told the author that the engines sounded a little like turboprops and another suggested that the sound was more like a steam train. It is the author's opinion that XA552, painted in its overall royal blue colour scheme, was perhaps the best-looking of all Javelins which, despite its aerodynamic weaknesses, was always an eye-catching type.

Rolls-Royce Avon

The second Mk.1 Javelin test bed was XA562 fitted with two Rolls-Royce Avon 209 units, an engine that during the 1950s was the direct competitor to the Armstrong Siddeley Sapphire. XA562 was allotted to Rolls-Royce at Hucknall for Avon RA.24R reheat development on 28 July 1955 (the RA.24R was an advanced Avon with reheat) and next day it was despatched to Rolls-Royce at Wymeswold (because at this time Rolls' Hucknall airfield was closed for runway construction). On 25 October it was taken to the Napier Flight Test Department by road, the conversion having been subcontracted to the Luton-based firm by Rolls-Royce.

So both the Avon and Gyron Junior Javelins were converted at Napier. For XA562 a very considerable degree of structural modification was required on the airframe to accommodate these physically bigger buried engines than the Sapphire, and to permit the use of afterburning. Together with the volume of special instrumentation required for the test programme, the work on XA562 made this the largest conversion so far undertaken by Napier, surpassing anything previously done (the task to fit XA552 later with Gyron Juniors proved to be an even bigger undertaking).

This test programme was part of the planning for the new supersonic English Electric Lightning fighter and XA562's twin RA.24R engines each had a Lightning reheat jetpipe with their variable nozzles operated by rams. At the time of XA562's arrival at Hucknall P.1B/Lightning reheat development was in the hands of English Electric Canberra WD959 (which XA562 would replace) and P.1B XA856. The Javelin was eventually delivered by road to Hucknall on 19 January 1958, but problems with reassembling the airframe delayed the first flight until 3 July, and little flying was achieved during the rest of that year.

XA562 was flown out of Hucknall and Wymeswold until it was grounded for a period in mid-1960, and then from June to October 1960 the aircraft completed 56 flights totalling 55 flight hours, of which 13 were in reheat. The maximum altitude at which the reheat was operated was 55,000ft, at which point it took 11 seconds to light up, and the flying programme was concerned mainly with engine handling when reheat was in operation, including ignition and flame stabilisation and the development in particular of the fully variable nozzle, plus component operation and reliability. These tests revealed that the reheat system stability and control was satisfactory on climbs from 20,000ft to 48,000ft and no extinction took place at the latter altitude, which at the time was the highest that the Javelin could operate because of oxygen limitations. The starboard jetpipe was subsequently modified to include a new pattern of variable nozzle with screw jack operation, in the form intended for the Avon RB146 (300-series) engine for the later Lightning F.Mk.3. This nozzle was fully variable in area.

On 9 August 1962 *Flight* magazine reported that 'considerable experience has now been acquired with RA.24R powerplants mounted in Javelin XA562 [and] a development programme on these engines has been in hand since September 1960 with an early P.1B (XA856)'. However, XA562's flying had come to an end some months earlier. On 28 February 1962 its starboard undercarriage refused to lower and the aircraft had to make a two-wheel landing at Hucknall. It swerved off the runway on to the grass and sustained minor damage. This was repaired but the programme was now discontinued. In all, XA562 had completed 182 flights with the Avons installed, totalling 174 hours. The workload was never intense but XA562 often had to fly flight profiles equivalent to those for the Lightning, i.e., up to 55,000ft with assessments of stability and reheat light-up.

No maximum performance figures were taken with the Avon Javelin since its purpose was to develop the reheat system, and so speed was not important. For anyone who might be interested to find out if changing an engine made any difference to a specific aircraft type, this situation applied in fact to just about all of the gas turbine test beds. Hunters and Lightnings etc., had structural (Mach) limitations, so fitting more powerful engines would have little effect on their speed because they were not allowed to fly any faster, and the extra power was used for heavier take-offs (higher weights) and to improve the climb performance. Few of the Hucknall flight test reports on gas turbine aircraft ever give sets of speed performance data, excepting examples of the Gloster Meteor, de Havilland Vampire and the American Lockheed Shooting Star.

On 14 August 1962 XA562 was allotted to the rocket testing site at Spadeadam in Cumberland for noise tests. The aircraft's outer wings were cut off, outboard of the wheels, and it was dismantled and despatched to Spadeadam by road via No 60 Maintenance Unit on 5 October 1962. XA562 was finally released from this role on 30 June 1965 and was Struck off Charge as scrap on 2 November 1967.

Plans were outlined to use further Javelin Mk.1s as test beds. XA560 was at one stage earmarked to have Rolls-Royce Conway engines installed, but in March 1955 this aircraft was allotted to Armstrong Siddeley for a trial installation of the Sapphire Sa.7; these later became development trials which effectively made it the Javelin FAW.Mk.7 prototype. XA564 was allotted to Bristol for Olympus engine development in connection with the Gloster Thin Wing Javelin programme, since this follow-on supersonic fighter was to have been powered by this engine. It arrived at Filton on 14 October 1955, but was then released from this role after the Thin Wing Javelin had been cancelled.

Above: Javelin XA562 newly converted with its twin Rolls-Royce Avon powerplant. Note the metal nose instead of a radome and the small faired bulges beneath the inner mainplane, which also appeared on Gyron Junior XA552 but not on standard production Javelins. Far fewer photos are available of this Javelin test bed. (Rolls-Royce Heritage Trust, Bristol)

Right: XA562 pictured at the Spadeadam test site where it was used for noise trials. (Peter Green)

This view of XA562 at Spadeadam shows how the outer wings had been removed before the aircraft began its final piece of trials work. Behind the aircraft is the former test tower used for the Blue Streak missile programme. (Peter Green)

Chapter 5
English Electric Canberra

Sapphire-Canberra WD933 pictured during a flypast at the Farnborough Airshow of 1954. (Terry Panopalis collection)

The jet bomber used most extensively as a test bed was the English Electric Canberra, and not just for engines. Weapons and radars and other avionics and equipment were all put on trial by this exceptionally versatile aeroplane. Canberra proved ideal for in-flight jet engine development with its docile handling, plenty of space for monitoring equipment, and it could operate at great height with the crew inside a pressured cabin. Aircraft previously used to test engines, Avro Lancasters and Lincolns (what *Aeroplane* magazine termed 'primary flying test beds'), could not offer the performance of types such as the Canberra and Meteor, which enabled pilots to extend both the altitudes and speeds attainable for test purposes. However, in several cases, these jet-powered machines were still hampered by the maximum value of engine thrust that their airframes could accept, which prevented combinations such as the Olympus-Canberra and Sapphire-Canberra in this chapter from flying in level flight at full throttle. (For more information on Canberra test beds the reader is advised to consult Dave Forster's excellent book *Black Box Canberras*.)

Nene Jet and Spectre Rocket
All of the original Canberra prototypes had Rolls-Royce Avon axial engines except one, VN813, which first flew on 9 November 1949 with 5,000lb Rolls-Royce Nene RNe2 centrifugal units. These were an 'insurance' against any problems that might delay the more advanced Avon (which at the time had difficulties with its compressor). Fortunately, production airframes got their Avons and so VN813 moved on to testing the de Havilland Spectre rocket engine.

The Nene's width had required bulged nacelles and there were no intake bullets. As a Canberra prototype VN813 spent time at its Warton birthplace and at Hucknall (for high-altitude trials), and then during 1951–52 with the Telecommunications Research Establishment (TRE) at Defford. A total of 180.4 hours was accumulated on Nene testing before, having been rejected for use as an Avon reheat test bed, on 8 June 1953 VN813 joined de Havilland Engines. The Spectre motor was under development for purely rocket-powered high-altitude interceptors (all of which were eventually

Above: The Rolls-Royce Nene-powered prototype Canberra VN813 captured on camera early in its career at Rolls-Royce Hucknall. The forward sections of the nacelles were made fatter to accommodate the Nene, but this change reduced the Canberra's ceiling. (Rolls-Royce)

Right: VN813 seen during ground testing of its new de Havilland Spectre rocket installation. (Rolls-Royce Heritage Trust, Filton)

This in-flight view of VN813 was taken during a hot firing test, possibly on 7 February 1957. Note the large tail bumper attached to the lower rear fuselage to protect the rocket barrels from hitting the runway when taking-off. However, this feature limited the aircraft's angle of attack on the ground, so it was not possible to fly from Hatfield with maximum fuel. Hence, these trials were made from the longer runway at RAE Bedford. (de Havilland Engines)

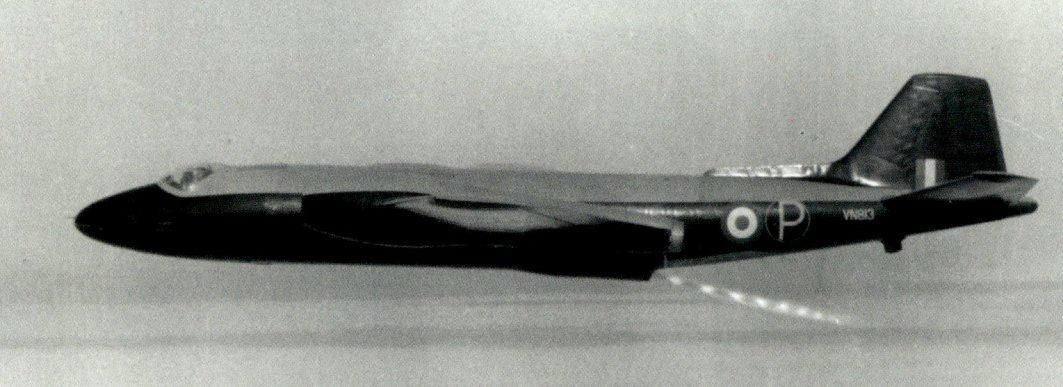

Above: For the 1957 Farnborough Show VN813 was painted bright blue with 'DE HAVILLAND Spectre CANBERRA' in red on the nose. (Rolls-Royce Heritage, Filton)

Left and below: These airborne shots of VN813 with its D.Spe.1 Spectre installed were taken in October 1957. (Rolls-Royce Heritage Trust, Filton)

The replacement for VN813 for the de Havilland Spectre trials programme was B.6 WJ755, which had a D.Spe.5 fitted. (Rolls-Royce Heritage Trust, Filton)

cancelled) and the Nene machine was picked simply because of a shortage of Canberras for trials, the bulged nacelles making it less suitable aerodynamically.

The initial structural work, including fitting a 600gal high-test peroxide (HTP) tank in the forward bomb bay, was undertaken by Folland at Chilbolton, VN813 arriving there on 24 June. It returned to Hatfield on 9 July 1954 to have a 8,000lb thrust Spectre D.Spe.1 installed in the remaining bomb bay, which was delayed until late 1956. Initial (very noisy) ground firings were made over a special pit, and the first airborne (cold) firing followed in mid-December. A 29-flight hot-firing programme began in February 1957 to assess the motor's performance and control, but the Spectre was found to have a quite low Mach number limit. In September 1957 de Havilland chief test pilot John Nicolson displayed VN813 at the Farnborough Show. The rear-bulkhead installation meant full power could not be used on take-off, but VN813's very rapid climb with the Spectre was something to remember! Finally, on 21 December 1959 VN813 was sold for scrap.

In the meantime, Canberra B.6 WJ755, built in 1954, went to de Havilland on 6 June 1957 as a replacement Spectre airframe. Folland again modified the aircraft (it arrived at Chilbolton on 15 July 1957 and returned to Hatfield on 24 April 1959) and during its 1959–60 trials with a D.Spe.5, without VN813's fatter nacelles, WJ755 was able to fly higher than the early prototype. After completing the D.Spe.5 programme in mid-1960, WJ755 undertook flight trials with an HTP-powered auxiliary power unit (APU) until 1962. On 8 February 1962 it was sold for spares recovery.

Avon and Reheat

On 5 October 1950 prototype VN850 went to Hucknall for Avon RA.3 trials, and then from mid-1951 it was re-engined with the more powerful 7,500lb RA.7 earmarked for the Hawker Hunter and Supermarine Swift. As such it first flew in early June, but then tragically crashed into Bulwell Common railway sidings on the 13th killing Rolls-Royce test pilot Richard Peach.

The replacement was B.2 WD930 completed in early 1951 and which reached Hucknall on 22 August. It began RA.7 testing in November, in particular assessing engine surge at between 38,000ft and 48,000ft and at 150 knots minimum. However, in early 1952 WD930 was reallotted to development of the RA.14 version, which now had greater priority, the conversion (which needed new nacelles of a different design) keeping the aircraft on the ground until first flight on 28 February 1953. The RA.14

This picture of Rolls-Royce Avon test bed WD930 was taken in 1956. (Rolls-Royce via Terry Panopalis)

gave more power, enabling WD930 to fly at 63,000ft and at the 1953 Farnborough Show to perform near-vertical climbs after take-off. Next WD930 was fitted with an RA.29 civil Avon in the port nacelle. It first flew on 29 August 1956 and attended the September SBAC Show. RA.29 trials ended in November 1957 and in all WD930 recorded 936.2 flight hours as an Avon test bed before being sold as scrap on 19 July 1961.

To replace WD930 and clear the RA.7 for production, brand new B.2 WF909 arrived at Hucknall on 18 July 1952. It flew with RA.7s in March 1953 and was used for general development work until 1956, before going to de Havilland (below). Another airframe used on Avon development was WH671, which arrived at Hucknall on 4 June 1954. It first flew with the RA.28 on 12 December 1955 and was sold as scrap on 1 May 1962.

The two-position clamshell nozzle for the reheated Avon RA.7R installation on WD943. (Rolls-Royce)

A further B.2, WD943, was allocated to development flying of the RA.7R, a version of the earlier Avon fitted with reheat. Having arrived at Hucknall on 17 October 1951, again brand new, it first flew with two RA.7Rs in place on 26 June 1952. The nacelle ends had been strengthened and enlarged to take the reheat pipe and there was a two-position clamshell nozzle. This extra power enabled WD943 to conduct reheat development at greater altitudes than the Meteor test beds could achieve in Chapter One. A visit to the Farnborough Show followed in September 1952, and then in 1957 WD943 moved to trialling different forms of reduction jet pipe nozzle. This Canberra was eventually sold as scrap on 20 November 1962 having completed 474.1 flight hours.

Finally, B.2 WD959 arrived at Hucknall on 21 November 1953 for RA.24 reheat development for the Lightning, but with an example installed only on the port side (an RA.7R in the starboard nacelle acted as a slave). With reheat engaged the RA.24 gave 14,370lb thrust and the trials flying continued from June 1957 until 1959. On 7 December 1959 WD959 was taken by road to No 12 School of Technical Training at Melksham as Ground Instruction Airframe 7620M. It was Struck off Charge on 7 October 1964.

The different geometry of the engine nacelle for the RA.7R can just be seen in this air-to-air image taken in September 1952. (Shell via Dave Forster)

Canberra B.2 WD959 was fitted with an experimental reheated Avon RA.24 in the port nacelle only, in a development programme for the supersonic Lightning jet fighter. This picture was taken at Wymeswold in February 1955. (Rolls-Royce)

Resurfacing the runway at nearby Hucknall meant that sections of the Rolls-Royce test fleet were based at RAF Wymeswold from January 1955 until February 1956. This view of WD959 was also taken in February 1955 and shows the bulged and stretched jetpipe. (Rolls-Royce)

Sapphire

On 13 April 1951 newly completed B.2 WD933 arrived at Bitteswell to serve as a test bed for Armstrong Siddeley's 7,200lb thrust Sa.3 Sapphire, and it made its first flight with these engines on 14 August. High-altitude trials began in January 1952, and then in April new 8,300lb Sa.6s were installed (the version earmarked for Hunters and Javelins). Extreme vibration was experienced early on which required a fix by English Electric, but testing through 1953 and 1954 brought performance flights at ever-higher altitudes until, by May 1954, the Sa.6s had taken WD933 to 56,000ft. This Canberra also attended the 1952 SBAC Show at Farnborough.

In 1954 10,200lb Sa.7s were fitted that made WD933 (in Armstrong Siddeley's words) the most powerful Canberra yet flown, which it did for the first time on 13 August. Farnborough followed in September, climbs to 53,000ft had been made by October, but then on 10 November WD933 crash landed at Bitteswell after one engine had failed to relight. Despite the aircraft flipping over the crew were okay, but WD933 was Struck off Charge on 30 November 1954. It was replaced at Bitteswell on 14 January 1955 by WK141, which first flew with Sa.7s installed on 7 May. By August the Canberra/Sapphire combination had been tested up to heights of 55,000ft and WK141 continued in this role until 1958, before moving on to the Viper engine (below).

To test fly Sapphires with reheat (Sa.6Rs), new B.2 WV787 arrived at Bitteswell on 1 September 1952, but the initial installation with fixed-area jet pipe nozzles took time to complete and delayed flight testing until March 1954. In June it was displayed at the National Air Races and this Canberra continued on the

Armstrong Siddeley Sapphire test bed WD933 taken during one of its visits to a Farnborough Show. (Armstrong Siddeley via Terry Panopalis)

The Canberra used to test Sapphires with reheat was WV787, seen here in mid-1954. (Mike Dowsing via Dave Forster)

reheat programme until the end of 1957. In 1958 it was offered for disposal from the Bitteswell Test Fleet and on 5 June went to Seighford to begin Blackburn NA.39 (Buccaneer) radar trials with Ferranti, having reverted back to an Avon powerplant. Today, WV787 survives with the Newark Air Museum.

Olympus

The Avro Vulcan's Bristol Olympus engine was also tested by the Canberra. On 18 December 1951 new build B.2 WD952 arrived at Filton to have two 9,750lb Olympus B.Ol.1/2B engines fitted in what proved to be a difficult adaptation. The Olympus units were larger, longer and heavier than WD952's original Avons, and its jetpipe would not fit within WD952's wing spar 'banjo' ring, so instead two derated 8,000lb units called Olympus 99s were installed to reduce the still-considerable structural changes. The additional weight and more forward position of the engines also necessitated some compensating ballasting.

On 5th August 1952 Bristol test pilot Wg Cdr Walter F. Gibb conducted WD952's first flight, from Filton, with Olympus 99s in place. In fact, this was the Olympus's maiden flight as well, all testing to date having taken place on the bench. The trials were most encouraging and it was soon clear that the aircraft's ceiling might possibly exceed the world altitude record. On 4 May 1953 an attempt was made

The Olympus Canberra WD952. In addition to its new engine installation, WD952 has a wingtip pitot.

Above: WD952 pictured in 1955 after having more advanced Olympus engines fitted. (Crown Copyright via Phil Butler)

Above: WD952 takes off for the flight that enabled the aircraft to break the world altitude record for a second time on 29 August 1955.

Below: In preparation for its next display flight, WD952 is towed away from the static display area at the SBAC Farnborough Show of September 1955. (Peter Berry)

Above left and above right: The second Olympus-Canberra test bed was WH713, which replaced WD952 in 1957.

Right: WH713 on view at the annual SBAC Show on 3 September 1957. (Peter Green)

on the record and a new official figure of 63,668ft was set. In the first half of 1954 WD952 was re-fitted with production-standard 11,000lb B.Ol.1/2C engines and made its first flight with them in June. The programme assessed the B.Ol.1/2C's performance, including relights, up to very high altitudes but (as noted in the opening paragraph) applying full power from these units was not permitted because it could damage the airframe. A trip to the Farnborough Show followed in September 1954 (this Olympus-Canberra had also attended the 1952 event and would do so again in 1955).

The following year 12,000lb B.Ol.11s were installed and the extra power enabled Gibb to break the altitude record for a second time on 29 August 1955 with a new figure of 65,876ft. However, on 12 March 1956 WD952's career came to an abrupt end when the port engine failed on take-off from Filton and it hit a tree over half a mile from the runway, which broke off the port wing; fortunately, the crew were unhurt. The remains were scrapped and WD952 was Struck off Charge on 18 January 1957.

To take its place WH713, first flown in early 1953, came to Filton in January 1957 to have 13,500lb thrust Olympus 104s fitted, though this installation did have its fuel flow restricted to reduce the thrust rating to 10,000lb, to match up with Bristol's new Zephyr commercial engine. The first flight with 104s took place in August 1957 and, for its appearance at the Farnborough Show a month later, WH713 featured fluted jet-nozzles to cut the levels of noise. The Olympus programme ended in mid-1959 and on 26 May 1960 WH713 was sold as scrap. By now sufficient Vulcans were available to undertake any necessary future Olympus trials.

Gyron Junior

After WF909 had finished its Avon work at Hucknall (above), on 8 December 1955 the Canberra went to de Havilland Engines at Hatfield to begin development flying with the 7,000lb D.GJ.1 Gyron Junior engine. At the start only one Gyron Junior was fitted, in the port nacelle, but this necessitated having lead ballast alongside to offset the new engine's greater thrust. The maiden flight arrived in late May 1957 but trials revealed problems with unsatisfactory compressor efficiency and with starting and control.

Above left: At the start of its career as a Gyron Junior test bed WF909 had only one engine installed, in the port nacelle. This view was apparently taken during the first flight in May 1957 and the difference between the Gyron Junior (nearest) and Avon nacelles is well shown. The fairing beneath the nearest nacelle may have been to hold ballast.

Above right: When initially converted for Gyron Junior trials WF909 retained its standard camouflage scheme.

Below: However, for the SBAC Show in 1957 WF909 was repainted in a bright blue colour scheme and had 'DE HAVILLAND Gyron Junior CANBERRA' applied in Day-Glo on the forward fuselage.

The Farnborough Show was attended in September 1957 and then, before year's end, a second Gyron Junior had been installed on the starboard side. The Gyron Junior was to power the upcoming Blackburn Buccaneer strike aircraft and improved versions (plus a Buccaneer intake on the port nacelle) were tested by WF909 until June 1962, when the programme came to an end. Released from task on 26 June, the Canberra was broken up at Hatfield.

Viper

Having completed its Sapphire testing, in September 1958 WK141 was reallocated to trials with Armstrong Siddeley's smaller Viper engine. In this role the Canberra replaced WK163 (below) and firstly a 1,750lb ASV.8, and afterwards a 2,460lb ASV.11 (as used by the Hunting Jet Provost), were installed and test flown inside a pod underneath the starboard wing outboard of the Avon nacelle. Following the closure of Armstrong Siddeley Motors at Bitteswell, on 18 September 1959 WK141 was despatched to Filton. After completing its flying career, WK141 was allotted for firefighting practice at Prestwick on 5 March 1963.

Right: Unlike all of the other Canberra turbojet test beds, WK141's involvement with the Armstrong Siddeley Viper engine saw the trials unit housed in an underwing pod, rather than as a replacement for the standard Avons. The reason was of course the Viper's small size. However, this aircraft also retained its Sapphire Sa.7 units from earlier trials and so was fully Armstrong Siddeley powered. (Terry Panopalis)

Below: Close-up of WK141's Viper ASV.8 installation in a picture dated 1 December 1958. (Bristol Siddeley Engines)

Scorpion Rocket

On 28 January 1955 brand new B.2 WK163 went to Bitteswell to conduct high altitude trials with a Viper ASV.5 installed in a pod under the starboard wing (initial plans to house the Viper in a wingtip pod had been dropped). This relatively short programme ran between April and August, after which the installation was removed to permit WK163 to join Napier for trials with that firm's Double Scorpion rocket motor. This was under consideration as an attachable pack for Lightning fighters to enable them to make interceptions at greater heights (B.2 WD929 had initially been allotted for the role until it was found that this aircraft needed some major repairs). Double Scorpion had two combustion chambers, each of them producing 2,000lb of thrust at sea level, and they could be fired independently so that pilots had either 2,000lb or 4,000lb on hand should they require it. The motor ran on kerosene fuel and used HTP as the oxidant.

WK163 arrived at Napier's Luton base on 2 December 1955 and was modified during the first months of 1956. Ground runs followed in April, though with only a single-barrel NScS.1 (100 Series) Scorpion. In addition, a portion of rear fuselage from retired Canberra VX165 was employed as a static test rig so that temperatures close to its surface could be measured during firing. The single-barrel form was taken aloft on WK163's first flight on 20 May 1956 crewed by Napier chief test pilot Mike Randrup and engineer Walter Shirley. By late May this had been run for 2 minutes at heights up to 30,000ft.

Soon afterwards the N.ScD.1-2 (200 Series) Double Scorpion was installed and the flying programme resumed during the second half of June, though the first flights still had only one combustion chamber operating. The Farnborough Show followed in September from where *Flight* magazine reported:

> Converted by the Napier Flight Development Establishment at Luton from a Canberra B.2, this test bed carries the Napier Scorpion liquid-propellant rocket motor at the rear of the bomb bay. The flying demonstration makes it evident that the Scorpion mounted in this aircraft has twin combustion chambers, one of which lights up a few seconds after the other. In action, the rocket gave evidence of HTP and hydrocarbon fuel, from the quality and colour of the flame and steam trail. It also showed itself to be acrobatically cleared.

Full Double Scorpion testing up to 40,000ft took place during October 1956.

The Lightning's 300 Series Double Scorpion pack came next and was fitted between February and May 1957. This motor, however, had more steeply angled barrels, which meant it needed to be further forward within the bomb bay to keep its line of thrust passing through the airframe CofG. As a result, the HTP tank was reduced to 350gal capacity, but the increased angle also meant the rocket flame was held further away from the fuselage surface, thereby permitting trials at far higher altitudes. Back in the air, WK163 performed successful hot-weather trials in Tripoli in June 1957 and then, back in the UK, in early August 1957 it climbed past 60,000ft for the first time. However, aircraft controllability deteriorated increasingly as the altitude went further above 50,000ft, and as the limiting Mach number and stall speed moved ever closer together. This situation was dealt with by attaching small vortex generators to the upper wings and to the fin, which now made flight possible to 70,000ft altitude (although over just a 15-knot speed range). This enabled Randrup and Shirley to attack the world altitude record on 28 August 1957 and to set a new figure (the third for a Canberra) of 70,310ft. In fact, Randrup and Shirley had already exceeded the old record several times during Double Scorpion test flights, passing 70,000ft on one occasion, and a record

Above left: Canberra WK163 served as a test bed for Napier's Double Scorpion rocket motor. (Napier)

Above right: The Double Scorpion was mounted at a downwards angle in the rear of WK163's bomb bay and inclined to ensure that, for stability reasons, the thrust line passed through the centre of gravity. The bomb bay doors were cut away in the region of the motor, a stainless-steel plate beneath the rear fuselage protected the skinning from the heat generated by the rocket, and a 450gal tank of high-test peroxide went in the front of the bomb bay. (Napier via Dave Forster)

Above left: The alterations to WK163's fuselage and the cut-back bomb bay doors are well shown in this view from June 1956. (Napier)

Above right: A few days after breaking the world altitude record WK163 attended the 1957 Farnborough Show with a green, red and yellow scorpion adorning each side of the nose. In addition, the legend 'WORLD AEROPLANE HEIGHT RECORD 21,350 METRES 70,000 FEET' had been added beneath 'NAPIER SCORPION CANBERRA' on the fuselage sides, just to the rear of the cockpit canopy. (Phil Butler)

Right: Superb photo of WK163 flying under normal Avon turbojet power but with the Double Scorpion ready to fire.

attempt made just before the official sortie had reached 73,000ft, but here the official barograph recorder had failed and so the figure was unfortunately invalid.

Double Scorpion trials continued until April 1958, the effort including support flights for fitting the motor in WT207 and WT208, a pair of B.6s earmarked for high-altitude nuclear sampling. However, on 9 April 1958 the engine in one of these RAF machines, WT207, exploded whilst under test at high altitude, which resulted in the survivor and WK163 being grounded. WK163 subsequently received a modified motor and appeared with it at the 1958 Farnborough Show. By now, however, the Lightning could reach very high altitudes on reheated Avons alone and the need for the Double Scorpion had passed. Test flying ended in October and the rocket motor programme was cancelled in 1959, although further unflown installations were assessed until March 1959.

With just the motor and HTP tank removed (in April 1959), but with the altered bomb-bay doors, bomb-bay fairing and vent pipe retained, WK163 was transferred to the RRE (Radar Research Establishment) at Pershore to undertaken new radar trials work. The surviving Scorpion fittings were finally taken away in 1966, and in 1994 the aircraft went on the civil register as G-BVWC (but still marked as WK163) for display flying. Today this famous Canberra is currently undergoing restoration to airworthy standard at Doncaster.

Chapter 6
Short S.A.4 Sperrin and Vickers Valiant

The Short S.A.4 Sperrin prototype VX158 with a Gyron installed in the lower engine nacelle. (Short Brothers)

The series of four medium-size nuclear bombers produced by the UK in the 1950s – the Valiant, Vulcan and Victor, which in service became the V-Force, and the Short Sperrin – did their share of test bed duties. The Victor itself, however, was never used as an engine test bed.

Short Sperrin

Built as a prototype bomber, but not followed by a production run, Sperrin VX158 stands apart from all of the other aircraft considered in this book. But once a Gyron unit had been installed the role of this particular airframe was changed and it had would now operate purely as a test bed. The production plans for the Sperrin were squashed by the arrival of the V-Bombers, all of which employed more advanced wing shapes than the straight form used by the Shorts bomber. However, the Sperrin design was itself sound, and the two prototypes would fulfil important careers in research flying.

Powered by four Rolls-Royce Avon engines, VX158 made its first flight from Short Brothers and Harland's Aldergrove airfield on 10 August 1951; the second Sperrin, VX161, made its first flight on 12 August 1952. Having completed its type trials, on 23 January 1954 VX158 was assigned to the 'fitting of special engines' – the new 15,000lb-class de Havilland Gyron designed specifically for supersonic fighters. The bomber's heavy structure and superimposed twin-engine nacelle arrangement meant it was well suited to this role, despite a certain amount of necessary localised strengthening required for the now bulged lower nacelle. With a maximum permissible Mach number of 0.85. VX158 could assess the Gyron's characteristics at the lower end of the speed range. Obviously, a supersonic mount would be needed to conduct tests at the high end of the range, but sadly that stage was never reached because the Gyron programme was to be cancelled.

Top: VX158 was the first Sperrin prototype and is seen here in original form with four-Avon powerplant. (Short Brothers)

Above: Side view showing VX158's original port engine nacelle arrangement with two Avon units inside. (Crown Copyright)

Right: By the September 1955 Farnborough Show VX158 had a de Havilland Gyron installed in the lower position of its port nacelle. (de Havilland Engines)

Above left: Close-up of the Gyron installation photographed on 31 August 1955. The Gyron's increased diameter had required the lower nacelle section to be widened and enlarged. (de Havilland Engines)

Above right: A Gyron display engine on exhibit at the 1955 or 1956 SBAC Farnborough Show. (de Havilland Engines)

Left: View of VX158 at the 1955 Farnborough Airshow with the Gyron in place. (Peter Berry)

Below: VX158 shows the difference in the lower nacelle size and shape for the Avon (starboard side) and Gyron (port). This picture was taken during a Farnborough display flight.

Another three-quarter in-flight view of VX158 taken when it had just one Gyron installed, but here with the undercarriage lowered.

The first phase saw a Gyron installed in VX158's lower port engine position only, the conversion being done in the Shorts' hanger at Aldergrove. A 25-hour special-category flight-approval test on the Gyron at 15,200lb thrust was conducted on the bench in January 1955, and then a 150-hour type-test schedule was successfully completed in August 1955 at the DGy.I's thrust rating of 15,000lb. By April 1956 the Gyron had completed a flight approval test at 18,000lb thrust and de Havilland could claim that this was the highest official rating of any turbojet in the western world. Also, back in January 1955 a 15,000lb engine had been boosted to 20,000lb by the addition of an afterburner, a fully variable convergent/divergent propelling nozzle was subsequently designed, and the maximum thrust in the Gyron DGy.2 was increased steadily to 20,000lb dry and over 25,000lb with reheat.

Back with the Sperrin, the first Gyron ground runs with VX158 took place on 6 July 1955 and the first flight followed on 7 July from Aldergrove with Short's test pilot 'Jock' Eassie and Chris D. Beaumont, de Havilland Engines' chief test pilot, in the cockpit. During the 30-minute flight the Gyron's operation proved very satisfactory, without any surge, and full rpm was achieved at approximately 10,000ft without trouble. By 21 July the maximum altitude to which the aircraft had been taken was 45,500ft and again the Gyron had behaved satisfactorily. However, on throttling back at 42,000ft to flight idling, the Gyron's flame was extinguished and it had proved necessary to return to base using VX158's remaining three Avon engines.

After the aircraft's preliminary handling trials had been conducted at Aldergrove, on 4 August VX158 was ferried to Hatfield for Beaumont to begin an intensive flight test programme with the engine's behaviour to be explored fully up to the limits imposed by the airframe. However, by 21 October 1955 engine relighting had proved satisfactory only up to altitudes of 20,000ft. A week later, slam accelerations, although possible, had been accompanied by what was thought to be transient compressor surging at altitudes of 20,000ft and upwards. VX158 attended the September 1955 Farnborough Show where *Flight* magazine noted how, with all three Avons idling, 'the Sperrin sailed by comfortably on the power of the single Gyron carried asymmetrically in its port nacelle'. And an indication of the acceleration now available was that the Gyron could change from flight idling to full thrust in only 3 seconds!

On 4 December VX158 returned to Aldergrove to have a second Gyron installed, this time in the lower starboard nacelle. The first ground run took place on 4 June 1956 and, with Eassie and Beaumont as crew again, VX158 made its first flight in this form on 26 June. However, during this sortie, when the aircraft was flying at 390 knots IAS (indicated airspeed) and 10,000ft, the port outer undercarriage door broke away and fell into the sea. Having been patched up and flown from Aldergrove to Queen's

Above: VX158 now with two Gyrons in position. From the front the Sperrin possessed an aggressive appearance. (Short Brothers)

Left: With both Gyrons installed the Sperrin makes a dramatic entry to the 1956 Farnborough Show. (Crown Copyright)

Flying view of VX158 with two Gyron units, taken most likely in August 1956.

Island in Belfast, in July VX158 was refitted with the equivalent door from VX161 in a move that brought the second Sperrin's flying career to an end.

After completing manufacturer's (Shorts) trials on the double-Gyron installation (including reaching the maximum airspeed in level flight of 390 knots IAS and the maximum g limit of 2.7), VX158 was flown back to Hatfield on 29 August 1956 by de Havilland test pilot Robert Plenderleith. In September the aircraft took part in its fourth Farnborough Show, the crowds experiencing the incredible noise generated by two Avons and two Gyrons running together on full power, one of the first occasions when such high decibel levels had been heard by the general public. This additional power also meant that the Sperrin's rate of climb at sea level was now much higher than when it had been powered solely by 6,000lb thrust Avons.

By 25 October successful relights had been achieved at heights up to 25,000ft, whilst on 7 November a 1 hour 40 minute-flight was devoted, in part, to acceleration tests with successful accelerations accomplished at heights up to 35,000ft, but at 40,000ft an interstage stall was experienced. Flights made in March 1957 included one on the 7th when Gyron performance calibrations were completed to the maximum test altitude of 45,000ft. Soon afterwards VX158 was grounded pending plans for further trials, but fate now took a hand.

In April 1957 the new Defence White Paper cancelled all future manned fighter programmes except the English Electric Lightning. As a by-product, this also brought termination for some fighter engine programmes including the Gyron, which meant the engine was never able to begin supersonic flights. Flight trials with VX158 did continue until the autumn of 1957, the research now including measuring the levels of infrared radiation produced by a large turbojet like the Gyron, the knowledge gained going towards the development of de Havilland's new heat-seeking air-to-air missiles. Further ground running tests had been completed by mid-January 1958, but by 15 May 1958 VX158 was up for disposal. In September 1959 it was sold to de Havilland and towards the end of that year the first and last Sperrin went for scrap.

Vickers Valiant

Two Vickers Valiants, second prototype WB215 and production aircraft XD872, were used in trials with de Havilland Super Sprite rocket motors to provide the bomber with additional thrust on take-off, the objective being to clear the rocket for service with the Valiant. In addition, the first production airframe WP199 was employed as a test bed for the Bristol BE.53 Pegasus vectored thrust turbofan engine, which was used by the Hawker/Hawker Siddeley P.1127 Harrier vertical take-off aircraft.

WP199 would spend all of its career on trials flying. After its maiden flight on 21 December 1953, it was used for Valiant clearance and armament testing, before spending time with A&AEE from 18 December 1954 as a trials aeroplane for airflow investigation and radar clearance. In August 1957 it joined Avro to test an inertial navigation system for the Blue Steel missile, and in September 1959 went to Martin Baker at Chalgrove for experiments with ejections seats for V-Bomber navigators. Then on 3 January 1961 it was flown to Bristol Siddeley at Filton in readiness for trials with the Pegasus.

The engine itself was bench tested for the first time in August 1959 while tethered hovering trials with P.1127 prototype XP831 had commenced in October 1960. The role for WP199, as a flying test bed with a Pegasus and its unique rotating exhausts installed in its bomb bay, was to assist in the development of this new and quite revolutionary engine. In fact, the engine's configuration meant that the installation was not entirely representative of that for the Harrier – the air intake for example was totally unlike that used by the Hawker aircraft.

Converting WP199 to take a 13,500lb thrust Pegasus 3 took time, but the first flight finally took place on 11 March 1963 with test pilot Tom Frost at the controls. Unfortunately, on flight three, and before much data had been gathered, the belly installation began to disintegrate after the nacelle that had covered the Pegasus intake began to peel back. The resulting damage halted the test programme for three months, while another delay came as a result of 'foreign object debris' being sucked into the Pegasus during a ground run. In the end, because of these and other problems, just 50 of the planned 70 test sorties were completed and the Valiant/Pegasus trials programme came to a close on 26 November 1964; nevertheless, useful data and experience had been acquired in handling a Pegasus in the air. On 15 March 1966 WP199 was sold for scrap and was Struck off Charge on 22 April.

Above and left: Vickers Valiant WP199 pictured flying with a Bristol Pegasus in position within the bomb bay and beneath the fuselage. (Rolls-Royce via Terry Panopalis)

Below: Poor quality but rare side view of WP199 serving as a Pegasus test bed. Unfortunately, the unique air intake for this installation does not show clearly. (Alan Constable via Terry Panopalis)

Chapter 7
Avro Vulcan

This slide photo is thought to show Vulcan prototype VX770 flying at the September 1958 Farnborough Airshow, with Rolls-Royce Conway engines installed. (Terry Panopalis collection)

The Avro Vulcan was powered by early versions of the Bristol Olympus turbojet, although first prototype VX770 made its original flights with four Rolls-Royce R.A.3 Avons and later with Armstrong Siddeley Sapphires. In all four Vulcans, including VX770, would test other jet engines and in the cases of the Rolls-Royce Spey and Conway the alternative units were mounted in the engine bays as direct replacements for the Olympus. However, for the more powerful Olympus 22R and 593 developed for the TSR.2 strike aircraft and Concorde supersonic airliner respectively, and for the Tornado's RB.199, the trial engines went in special pods mounted inside and beneath the centre fuselage bomb bay.

Conway 1

The first prototype Avro 698 (then still unnamed), VX770 with its Avon engines, made its maiden flight on 30 August 1952. With the Olympus still not ready, in 1953 this aircraft had more powerful 8,000lb thrust Sa.6 Sapphires installed to enable Avro to conduct test flying at the highest maximum indicated airspeed and maximum Mach number, something not possible with the Avons. Flying resumed in July 1953 and on the 15th of that month VX770 was displayed at the RAF Review held at Odiham to mark Queen Elizabeth's coronation. Here, for the first time, Avro's Roly Falk was able to display the bomber at speed, arriving at the show at roughly 460mph. Ten days later VX770 put in an appearance at a show held at Royal Naval Air Station Stretton. The prototype was Accepted off Contract at Avro on 18 January 1955 but it was not until 1956 that VX770 became involved with the Rolls-Royce Conway.

Above and below left: These views of VX770 were made at the 1954 Farnborough Airshow when the aircraft was powered by Sapphire engines. (colour – Terry Panopalis collection)

Below right: Vulcan prototype VX770 pictured while serving as the Rolls-Royce Conway flight test bed. (Rolls-Royce via Terry Panopalis)

Proposals were made to fit Conways in a Handley Page Victor B.1 for trials, either under the wings or alternatively in a pod in the bomb bay, since the Victor's intakes were not sufficiently large to supply air to two Conways in the normal engine positions, but these were rejected. The Vulcan's air intakes were large enough, however, and this brought VX770 into the picture, although the prototype was of course not representative of a production type and so there were stringent performance limits to avoid exceeding the airframe limits.

VX770 was delivered to Avro's Langar facility for conversion on 17 August 1956; initially it had Conway R.Co.7 engines installed and first flew in this form on 9 August 1957. In fact, the Conways had required larger intakes and jetpipes within the Vulcan airframe, but VX770's wing planform was unchanged. On 24 August the Vulcan was delivered to Rolls-Royce's Flight Test Establishment at Hucknall to begin trials looking at the engine's handing and reliability. The aircraft also took part in the 1957 Farnborough Show, and then during November and December the Co.7 trials were completed in Malta to avoid delays caused by bad weather.

Afterwards VX770 was grounded to be refitted with 17,250lb Conway R.Co.11 units for another intensive flight development programme and as such flew again in May 1958. The trials primarily involved taking readings during cruising flight to accumulate data for the engine in readiness for service in other aircraft types. VX770 attended the September 1958 Farnborough Show but because of its development engagements the 'Conway-Vulcan', as it was now called, was only able to make flypasts during the event and did not land.

Having accumulated 833 flying hours, on 20 September 1958 VX770 was destroyed in a tragic accident during the Battle of Britain display at RAF Syerston. During a low-level pass the starboard wing failed at around mid-wing and then broke apart, VX770 dived into the ground and all of the crew (Rolls-Royce test pilot Keith Roland Sturt, co-pilot Ronald W. Ward from Fairey, Rolls-Royce flight engineer William E. Howkins and RAF navigator Flt Lt Raymond M. Parrott) and three RAF staff on the ground died. The wing structure had failed because the airframe had been over-stressed and this loss would hold back the Conway test programme by 10 months.

TSR.2 Test Bed

The very advanced Olympus Ol.22R was developed to power the supersonic British Aircraft Corporation (BAC) TSR.2 strike aircraft. After its maiden flight on 9 January 1957, Vulcan B.1 XA894 spent considerable time on de-icing and autopilot trials with the Ministry of Supply Air Fleet at Woodford, and then from October 1958 with A&AEE for trials with the autopilot and navigational bombing system. Next it had an Ol.22R installed inside a nacelle positioned beneath the fuselage. This had bifurcated intakes geometrically similar to the TSR.2 configuration while the engine body was enclosed within a sealed sleeve, inside which boundary layer air provided cooling and fire protection.

XA894 arrived at Filton for conversion on 18 July 1960 and the first flight with the Olympus 22R took place on 23 February 1962. After accumulating nearly 80 hours of flight time, on 8 December 1962 XA894 was completely destroyed on the ground at Filton. The aircraft caught fire following an uncontained turbine failure during a ground run. In fact all five engines were running when the test crew heard a bang, fire immediately broke out and it eventually consumed the aircraft entirely (and one of the fire tenders which had rushed to the scene). Afterwards a turbine disc was discovered some 100 yards from the scene.

Right: Air-to-air view of XA894 with the Olympus 22R installation. The engine itself was situated partially within the bomb bay. (Bristol Siddeley Engines)

Below: This angle shows nicely the 'bifurcated' air intakes for XA894's TSR.2 engine. (Terry Panopalis collection)

Above: A view of the Olympus 22R Vulcan XA894 taken during a display flight at the 1962 Farnborough Show. The Vulcan engine test beds seem to have been regular performers at Farnborough. (Bristol Siddeley Engines)

Left: Another Farnborough 1962 view, this time showing airbrakes deployed and XA894's Olympus 22R with reheat lit. (Rolls-Royce via Terry Panopalis)

XA896

B.1 XA896 was allotted to serve as a test bed for the Bristol Siddeley BS.100 turbofan lift/thrust engine earmarked for the supersonic Hawker Siddeley P.1154 vertical take-off aircraft. The Vulcan arrived at Hucknall on 25 May 1964 to begin its conversion on behalf of Bristol Siddeley, with the engine to go in a pod beneath the fuselage. However, the P.1154 project was cancelled and in February 1965 work on the conversion was halted. XA896 was subsequently scrapped.

Conway 2 and Spey

With the loss of VX770, in late 1958 B.1 XA902 was allotted as a replacement to continue and complete the Conway programme (this Vulcan had first flown on 13 April 1957). The endurance tests had to be completed to enable the new engine to operate in Boeing 707 airliners and the Victor B.2. From 3 December 1958 XA902 had R.Co.11 engines fitted at Woodford, it first flew with them on 14 July 1959 and then arrived at Rolls-Royce Hucknall on the 17th.

    Apart from normal handling and relighting checks, the main task was to achieve a Conway overhaul life of 1,000 hours by flying as much as possible. Flights of six hours were made at a cruising altitude of

Above left and above right: Vulcan B.1 XA902, fitted with four Conways, performs at Farnborough in September 1959. (Terry Panopalis collection)

Right: XA902 was the Conway and Spey test bed and is seen here undergoing ground testing. When trial engines were housed in the Vulcan's normal Olympus positions it was often difficult to tell which engines were in place when a picture was taken. A date for the photo thus becomes very important. In fact, with two Speys installed, XA902 was indistinguishable from when it had four Conways. (Rolls-Royce via Terry Panopalis)

Spectacular view of XA902 coming into land at Rolls-Royce's Hucknall airfield. (Rolls-Royce)

40,000ft, but at this height the cruise rpm for the Conway meant that the Vulcan's maximum speed would be exceeded, thereby taking this aircraft beyond its airframe limits. Consequently, the Conways were operated at Boeing 707 airliner ratings, slightly lower than that for an R.Co.11, with two of the engines running at high rpm (9,370rpm) and the other two at a lower figure. In fact, this structure/power limitation problem had been responsible for VX770 breaking up in the air, because the prototype's structure was nothing like as strong as that in a production airframe, with the Mach limit correspondingly lower.

With four Conways in position XA902's maximum all-up-weight limit was 160,000lb, the same as a standard B.1. In January 1960 *Flight* recorded that XA902 had quite frequently been flying 18 hours in a day, and in total the Vulcan would compile 1,021 hours airtime in its all-Conway configuration. Much of the programme was flown from Hucknall, but between 5 January and 17 March 1960 'hot' weather sorties were made from the Rolls-Royce Engine Development Section based at Luqa in Malta.

With the Conway programme out of the way XA902 was modified to take the new Rolls-Royce RB.163 Spey engine, which would power the Blackburn Buccaneer naval strike aircraft and Hawker Siddeley Trident airliner. Speys were installed in the inboard engine positions only with the Conways retained in the outer bays. It made its first flight in this latest form from Hucknall on 12 October 1961 and after two flights company chief test pilot Jim Heyworth reported favourably on the new engine's handling qualities. The Spey flight test programme occupied just 41.5 flying hours over a period of four weeks, the main objective being to check a new all-speed engine control system in flight (all other Spey testing was performed on ground test beds and in rigs). Plans to install Rolls-Royce RB.141 Medway engines after the Spey programme were not followed up.

XA902 was finally Struck off Charge on 17 October 1962 and returned to the RAF for disposal. Over the following months it was carefully dismantled and on 7 March 1963 despatched by No 60 Maintenance Unit by road to RAF Dishforth. This final exercise was a feasibility study to see if it would be practical to transport Vulcan airframes in such a manner, should the need arise.

Concorde and Tornado

The BAC/Aérospatiale Concorde Olympus 593 engine, developed from the Olympus 22R above, was tested by Vulcan XA903, which had first flown on 10 May 1957. Prior to its Concorde role this aircraft had from May 1957 joined the Ministry of Supply Air Fleet at Woodford for Blue Steel missile development trials.

The box-shaped Olympus 593 nacelle beneath XA903 represented the Concorde installation.

Above: The Olympus 593 test bed performing at the September 1968 Farnborough. The conversion did not just involve fitting the nacelle – the bomb bay now contained water and fuel plus avionics and instrumentation for the trials programme. The bomb aimer's blister and the tail cone were removed. (Terry Panopalis collection)

Right: Underside view of XA903 with its Concorde engine. Note the deployed airbrakes. (Terry Panopalis)

Below: Concorde Olympus XA903 with the special spray rig and piping now fitted beneath the forward fuselage for water and ice ingestion tests. Water could be sprayed in at different concentrations with the nacelle's de-icing system running. A tail cone has been refitted and the first flight with the grid installed was made on 12 March 1971. (Bristol Siddeley via Terry Panopalis)

XA903 arrived at Filton for its Olympus 593 conversion on 3 January 1964. As before, the new power unit was installed underneath the fuselage, but here in the form of a Concorde engine pod with a single 'straight-through' intake. In this configuration the Olympus 593 thrust line was a full 7ft below that of the bomber's standard basic Olympus units. The maiden flight with the extra engine took place on 9 September 1966 and the trials embraced 219 flights totalling 417 hours of flight time, 248 hours of Olympus 593 running, another 109 hours of windmilling and 24 hours of ground running. *Flight* magazine reported that a total of 2,665 light-ups, 18 hours of reheat testing, and 1,279 reheat light-ups were also recorded. At one stage a water spray rig was fitted under the Vulcan's forward fuselage for water and ice ingestion trials.

A key element in the programme involved landing the Vulcan repeatedly at above its maximum design landing weight, and it was flown beyond its service limit speed. Remarkably, the handling qualities of the Vulcan/593 combination revealed that no significant pitch-trim changes were induced over the complete throttle range of the Concorde powerplant. The 593 programme ended in 1971 and over this five-year period six standards of engine and three intake configurations were flown in ten phases of testing. The final phase saw the installation of a -4 engine representative of the units installed in the first pre-production Concorde 01. The Vulcan test bed proved the ideal tool to assess the Olympus 593, in part because of the ease in which flight conditions could be changed.

The next engine tested by XA903 was the Rolls-Royce/Turbo-Union RB.199 developed for the Panavia Tornado Multi-Role Combat Aircraft and again the supplementary engine went in a pod below the fuselage. Here, however, the pod conformed to the starboard side of a Tornado fuselage and in due course would see the further fitting of a Tornado gun pod with its 27mm Mauser cannon. This would enable engine relighting, windmilling and reheat lighting to be carried out in a realistic environment. On 4 August 1971 XA903 was flown from Filton to Cambridge for Marshalls of Cambridge to receive its conversion. On completion it was delivered to Rolls-Royce on 9 February 1972 and then first flew with the RB.199 in position on 19 April 1973. A particularly unique feature of the programme happened in January 1976 when XA903 was flown to A&AEE Boscombe Down for firing trials with the Mauser cannon. These were performed on the ground, but the object was to determine whether ingesting the products from shell propellants might affect the engine's performance. In all, 285 flying hours were accumulated with the RB.199 before XA903, as the last Vulcan B.1 to fly, arrived at RAE Farnborough on 22 February 1979 for retirement as a ground training airframe. It was Struck off Charge on 19 July 1979 and today its nose survives at the Wellesbourne Wartime Museum.

The Rolls-Royce/Turbo-Union RB.199 mounted in its 'Tornado Fuselage' housing on the underside of XA903. This photo was taken at the September 1974 Farnborough Airshow.

RB.199 flight test bed XA903 seen taxiing. (Rolls-Royce)

Vulcan XA903 pictured as it was just about to make the final landing of its career, at RAE Farnborough on 1 March 1979. (RAE via Terry Panopalis)

Glossary and Bibliography

A&AEE	Aeroplane and Armament Experimental Establishment, Boscombe Down.
CofG	Centre of gravity.
IAS	Indicated airspeed.
MAP	Ministry of Aircraft Production – created in May 1940 to relieve the Air Ministry of its role of procuring aircraft and the equipment and supplies associated with them. Its functions were transferred to the Ministry of Supply in 1946.
Metro-Vick	Metropolitan-Vickers.
MoS	Ministry of Supply – created in August 1939 to provide stores used by the RAF.
NGTE	National Gas Turbine Establishment, Pystock and Bitteswell.
RAE	Royal Aircraft Establishment, Bedford and Farnborough.
SBAC	Society of British Aircraft Constructors.

Select Bibliography

Research for this book involved accessing a considerable volume of original documents, such as flight test reports, held in the British National Archives 'Air' and 'Avia' files and in various other collections of papers. Documents held by BAE Systems Heritage at Farnborough, the Farnborough Air Sciences Trust, Brooklands Museum, and the Rolls-Royce Heritage Trust at Derby and Filton, provided vital contributions, while the *Flight* Global Archive furnished some splendid 'on the spot' reports. The following secondary sources were also important:

Ashwood, P. F. & Lean, D., 'Flight Tests of a Meteor Aeroplane fitted with Jet Deflection', *Journal of the Royal Aeronautical Society,* London (August 1958)

Birch, David, *Hucknall - The Rolls-Royce Flight Test Establishment,* The Rolls-Royce Heritage Trust, Derby (2017)

Butler, Phil, 'Gloster Javelin Engine Test Beds' in *Aeromilitaria*, Staplefield (Summer 2011)

Butler, Phil, 'More Canberras - The Engine Test Beds' in *Aeromilitaria*, Staplefield (Spring 2007)

Butler, Phil, & Buttler, Tony, *Avro Vulcan: Britain's Famous Delta-Wing Bomber,* Aerofax/Midland/Ian Allan, Hinckley and Shepperton (2007)

Forster, Dave, *Black Box Canberras: British Test and Trials Canberras 1951-1994,* Hikoki, Manchester (2016)

Lindsay, Roger, *Service History of the Gloster Javelin Marks 1 to 6,* Self-Publication (1975)

Morgan, Eric B., *Vickers Valiant: The First of the V-Bombers*, Aerofax/Midland/Ian Allan, Hinckley and Shepperton (2002)

Sargent, David, '70th Anniversary of Time to Height Record by Sapphire Meteor F.Mk.8 WA820 - August 31st, 1951', *Journal of the Rolls-Royce Heritage Trust* (December 2021)